APPLICATIONS OF ARTIFICIAL INTELLIGENCE

MATTHEW N. O. SADIKU

Regents Professor Emeritus & IEEE Life Fellow
Department of Electrical & Computer Engineering
Email: sadiku@ieee.org or matthew_sadiku@yahoo.com
Web: www.matthew-sadiku.com

SARHAN M. MUSA

Professor
Department of Electrical & Computer Engineering
Prairie View A & M University
Prairie View, TX 77446

SUDARSHAN R. NELATURY

Associate Professor
Department of Electrical & Computer Engineering
Penn State, Erie

Gotham Books
30 N Gould St.
Ste. 20820, Sheridan, WY 82801
https://gothambooksinc.com/
Phone: 1 (307) 464-7800

Published by Gotham Books (June 1, 2022)

ISBN: 978-1-956349-78-8 (H)
ISBN: 978-1-956349-76-4 (P)
ISBN: 978-1-956349-77-1 (E)

BRIEF
TABLE OF CONTENTS

DEDICATION

TO OUR CHILDREN

Motunrayo, Ann, and Joyce

Mahmoud, Ibrahim, and Khalid

Charles Finney and Charles Wesley

PREFACE

Recently, we have witnessed a wave of emerging technologies, from the Internet of things and blockchain to artificial intelligence, demonstrating significant potential to transform and disrupt multiple sectors. Artificial intelligence (AI) refers to intelligence demonstrated by machines, while natural intelligence is the intelligence displayed by humans and animals. AI is an umbrella term John McCarthy, a computer scientist, coined in 1955 and defined as "The science and engineering of intelligent machines." AI is the development of computer systems that are able to perform tasks that would require human intelligence. From ancient times, humans have been dreaming of creating artificial intelligence. AI represents the hopes and fears of an industry seeking more intelligent solutions. It has the potential to change everything.

Typically, AI systems demonstrate at least some of the following human behaviors: planning, learning, reasoning, problem-solving, knowledge representation, perception, speech recognition, decision-making, language translation, motion, manipulation, intelligence, and creativity. AI is an interdisciplinary and comprehensive field covering numerous areas such as computer science, psychology, linguistics, philosophy, neurosciences, cognitive science, thinking science, information science, system science, and biological science. Today, AI is integrated into our daily lives in several forms, such as personal assistants, automated mass transportation, aviation, computer gaming, facial recognition at passport control, voice recognition on virtual assistants, driverless cars, companion robots, etc.

Although AI is a branch of computer science, there is hardly any field that is unaffected by this technology. Common areas of applications include agriculture, business, law enforcement, oil and gas, banking and finance, education, transportation, healthcare, engineering, automobiles, entertainment, manufacturing, speech and text recognition, facial analysis, telecommunications, and military.

The book is divided into nineteen chapters. Chapter 1 provides a comprehensive introduction to artificial intelligence and its components: expert systems, fuzzy logic, artificial neural network, and machine learning. deep learning, natural language processing, and robots.

Chapter 2 examines the role of AI in the different sectors of smart cities. Today, the focus is on smart cities, which aspire to connect all aspects of urban life.

While several technologies are needed to realize the concept of smart cities, the two key enablers of the smart city are the Internet of things and artificial intelligence.

Chapter 3 investigates the role of AI in the different sectors of the smart grid. The smart grid has emerged as a replacement for ill-suited traditional power grids designed as one-directional systems. It is an electrical power grid that is integrated with an AI-enabled, two-way communication network providing energy and information.

Chapter 4 provides an overview of a broad range of applications of AI in healthcare. Healthcare is regarded as the next domain to be revolutionized by AI. In healthcare, AI can help manage and analyze data. AI can have a significant impact in making healthcare more accessible.

Chapter 5 explores the advances of artificial intelligence applied in engineering. AI engineering is essentially the use of AI technologies in the development of AI applications. The field aims to equip practitioners to ensure human needs are translated into understandable, ethical, and trustworthy AI.

Chapter 6 deals with various applications of AI in education. Applications of AI in education are on the rise and are receiving a lot of attention at all levels of education. The promise of AI applications lies partly in their efficiency and efficacy.

Chapter 7 addresses various applications of AI in business. Businesses of all types and sizes are considering artificial intelligence to solve their problems. AI can increase productivity, gain competitive advantage, complement human intelligence, and reduce the cost of operations.

In Chapter 8, we cover various applications of artificial intelligence in the industry. AI has impacted our lives significantly by fostering advances in many industries such as healthcare, eCommerce, pharmaceutical industry, energy industry, agriculture industry, petroleum industry, telecommunications, industry, construction industry, online advertising, consumer electronics, education, and entertainment.

Chapter 9 addresses the various uses of artificial intelligence in the manufacturing industry. AI in manufacturing is a game-changer. AI is already transforming manufacturing in several ways and also changing the way we design products. Manufacturing has a strong association with AI, especially the use of automation. Today industrial leaders such as Google, Microsoft, Procter & Gamble, and IBM have invested heavily in AI.

Chapter 10 gives various applications of AI tools in agriculture. AI is transforming agriculture in many ways. Farmers are relying on AI technology in their crop production. Some companies are leveraging computer vision and deep learning algorithms to process the data captured by drones. Food producers are using AI to sort products and reduce labor.

Chapter 11 presents applications of AI in the food and beverage industry. AI is poised to revolutionize the food industry. AI technology can be implemented in all stages of the food supply chain, resulting in an overall improvement and a significant increase in efficiency. The application of AI in the food and beverage industry is already transforming the way we think about food production, food manufacturing, food processing, food quality, food delivery, food consumption, and food storage.

Chapter 12 deals with various uses of AI technology in autonomous vehicles. AI performs several tasks in a self-driving automobile. AI helps vehicles to navigate through the traffic and handle complex situations. AI can respond quickly to data points generated from hundreds of different sensors. Car manufacturers are using AI tools in every aspect of the car-making process.

Chapter 13 provides an introduction to artificial intelligence-based chatbots. At the heart of chatbot lies natural language processing, a branch of AI. An AI chatbot is usually trained to operate more or less on its own. It has a key advantage of being able to learn a lot about its users.

Chapter 14 explores the impact of various artificial intelligence tools on social media companies. AI is a key component of the popular social networks we use every day. In the digital age, AI is constantly transforming social media.

Chapter 15 covers how AI can be used in games in various ways. AI tools are used different aspects of a game to mimic, imitate, learn, forget, teach, and collaborate. AI techniques can help generate intelligent, responsive behavior that molds on your reactions as a player. AI makes the game more interactive by boosting player's experience.

In chapter 16, various uses of humanized AI are covered. Humanized AI is that which understands human emotions like happiness, stress, urgency, anger, to detect emotions like laughter, anger, arousal, and pain. It responds to natural language very much like a human friend. Humanizing principles can be applied to every machine that involves human-AI collaboration.

Chapter 17 focuses on the various applications of wearable AI. AI in wearables is aimed at improving the functionalities and user experience to provide users with real-time insights. Wearable AI gadgets such as smartwatches and fitness bands can be used to monitor health-oriented vitals such as heart rate and blood pressure. These gadgets are a boon for firefighters in critical rescue operations.

Chapter 18 introduces the use of artificial intelligence in cybersecurity. AI tools have been increasingly applied for cybercrime detection and prevention. They can be used to learn how to enable security experts to understand the cyber environment in order to detect abnormalities. They can also be used to help broaden the horizons of existing cyber security solutions.

Chapter 19 is the last chapter, which examines various applications of AI in the military and defense. The promise of AI (automation, informed decision making, self-control, self-regulation, and self-actuation of combat systems, etc.) are driving militaries around the world to accelerate research and development. The modern uses of AI in the military are not limited to the battlefields. AI can help reduced the risk of life loss in wars. AI can be used for training systems.

The second section of each chapter (with the exception of chapter one) provides an overview of AI. This may seem repetitious, but it is intended to make the chapter self-contained. If it makes reading grievous, one might remember Zig Zigler's quote, Repetition is the mother of learning, the father of action, which makes it the architect of accomplishment."

This book is designed to help learners decode the mystery of artificial intelligence and its various applications. It provides researchers, students, and professionals a comprehensive introduction, benefits, and challenges for each application area of AI technologies. The authors were motivated to write this book partly due to the lack of a single source of reference on the various applications of AI tools. Hence, the book will help a beginner to have an introductory knowledge about AI and its applications. The main objective of the authors is to provide a concise treatment that is easily digestible for each application. It is hoped that the book will be useful to practicing engineers, computer scientists, and information business managers.

DETAILED
TABLE OF CONTENTS

Chapter 9 - AI in Manufacturing

9.1 Introduction

9.2 Review on Artificial Intelligence

9.3 AI in Manufacturing

9.4 Applications of AI in Manufacturing

9.5 Benefits

9.6 Challenges

9.7 Global AI in Manufacturing

9.8 Future of AI in Manufacturing

9.9 Conclusion
References

Chapter 10 - AI in Agriculture

10.1 Introduction

10.2 Review on Artificial Intelligence

10.3 AI in Agriculture

10.4 Machine Learning in Agriculture

10.5 Applications of AI in Agriculture

10.6 Benefits

10.7 Challenges

10.8 Global AI in Agriculture

10.9 Future of AI in Agriculture

10.10 Conclusion
References

Chapter 11 - AI in Food Industry

11.1 Introduction

11.2 Review on Artificial Intelligence

11.3 AI Research Projects in Food Industry

11.4 Applications of AI in the Food Industry

11.5 Benefits

11.6 Challenges

11.7 Global AI in Food Industry

11.8 Future of AI in Industry

11.9 Conclusion

References

Chapter 18 - AI in Cybersecurity

18.1 Introduction

18.2 Review on Artificial Intelligence

18.3 What is Cybersecurity?

18.4 Applications of AI in Cybersecurity

18.5 Benefits

18.6 Challenges

18.7 Global AI in Cybersecurity

18.8 Future of AI in Cybersecurity

18.9 Conclusion
References

Chapter 19 - AI in Military

19.1 Introduction
19.2 Review on Artificial Intelligence

19.3 Military

19.4 Military AI

19.5 Applications of Military AI

19.6 Benefits

19.7 Challenges

19.8 Global AI in Military

19.9 Future of Military AI

19.10 Conclusion
References

ABOUT THE AUTHORS

Matthew N. O. Sadiku received his B. Sc. degree in 1978 from Ahmadu Bello University, Zaria, Nigeria, and his M.Sc. and Ph.D. degrees from Tennessee Technological University, Cookeville, TN in 1982 and 1984 respectively. From 1984 to 1988, he was an assistant professor at Florida Atlantic University, Boca Raton, FL, where he did graduate work in computer science. From 1988 to 2000, he was at Temple University, Philadelphia, PA, where he became a full professor. From 2000 to 2002, he was with Lucent/Avaya, Holmdel, NJ as a system engineer and with Boeing Satellite Systems, Los Angeles, CA as a senior scientist. He is presently a professor emeritus of electrical and computer engineering at Prairie View A& M University, Prair View, TX.

He is the author of over 970 professional papers and over 95 books including Elements of Electromagnetics (Oxford University Press, 7th ed., 2018), Fundamentals of Electric Circuits (McGraw-Hill, 7 the ed., 2021, with C. Alexander), Computational Electromagnetics with MATLAB (CRC Press, 4th ed., 2019), Principles of Modern Communication Systems (Cambridge University Press, 2017, with S. O. Agbo), and Emerging Internet-based Technologies (CRC Press, 2019). In addition to the engineering books, he has written Christian books including Secrets of Successful Marriages, How to Discover God's Will for Your Life, and commentaries on all the books of the New Testament Bible. Some of his books have been translated into French, Korean, Chinese, Italian, Portuguese, and Spanish.

He was the recipient of the 2000 McGraw-Hill/Jacob Millman Award for outstanding contributions in the field of electrical engineering. He was also the recipient of Regents Professor award for 2012-2013 by the Texas A& M University System. He is a registered professional engineer and a fellow of the Institute of Electrical and Electronics

Engineers (IEEE) "for contributions to computational electromagnetics and engineering education." He was the IEEE Region 2 Student Activities Committee Chairman.

He was an associate editor for IEEE Transactions on Education. He is also a member of the Association for Computing Machinery (ACM) and the American Society of Engineering Education (ASEE).

His current research interests are in the areas of computational electromagnetic, computer networks, and engineering education. His works can be found in his autobiography, My Life and Work (Trafford Publishing, 2017) or his website: www.matthew-sadiku.com. He currently resides Hockley, Texas. He can be reached via email at sadiku@ieee.org.

Sarhan M. Musa is a professor in Electrical and Computer Engineering Department at Prairie View A& M University. He holds a Ph.D. in Electrical Engineering from the City University of New York. He is the founder and director of Prairie View Networking Academy (PVNA), Texas. He is LTD Sprint and Boeing Welliver Fellow. Professor Musa is internationally known through his research, scholarly work, and his published books. He has had a number of invited talks at international conferences. He has received
a number of prestigious national and university awards and research grants. He is a senior member of the IEEE and has also served as the member of technical program committee and steering committee for a number of major journals and conferences. Professor Musa has written more than a dozen books on various areas in Electrical and Computer Engineering. His current research interests cover many topics in the artificial intelligence/ML, data analytics, Internet of things, wireless network, data center protocols, and computational methods.

Sudarshan R. Nelatury is an associate professor of electrical and computer engineering at Penn State University, USA. He received his M.E. and Ph.D. in electrical engineering from Osmania University (OU) in 1983 and 1996 respectively. He was with the Department of ECE at the University College of Engineering, OU during 1983-1999. He was invited by Villanova University in 1999 and worked as a visiting faculty till May 2003. He then moved to Penn State University in June 2003 and has been working in the electrical and computer engineering department till date. Nelatury authored/co-authored over 150 technical articles including 2 textbooks and 6 solutions manuals. During 2012, he was invited by the Moore School of Electrical Engineering, University of Pennsylvania (UPenn), where he spent his first Sabbatical. He was recipient of Outstanding Research Award and Council of Fellows Research Award in 2008 from Penn State Erie, The Behrend College for his prolific publication record. He has been a Senior Member of IEEE and was Life Member of IETE and ISTE of Indi

CHAPTER 1
INTRODUCTION

"If the government regulates against use of drones or stem cells or artificial intelligence, all that means is that the work and the research leave the borders of that country and go someplace else." —*Peter Diamandis*

1.1   INTRODUCTION

Artificial intelligence refers to intelligence demonstrated by machines, while natural intelligence is the intelligence displayed by humans and animals. It is the development of computer systems that are able to perform tasks that would require human intelligence. The field of artificial intelligence (AI) aims at creating and studying software or hardware systems with a general intelligence similar to, or greater than, that of human beings. AI is now one of the most important global issues of the 21st century.

Artificial intelligence is the branch of computer science that deals with designing intelligent computer systems that mimic human intelligence. Typically, AI systems demonstrate at least some of the following human behaviors: planning, learning, reasoning, problem solving, knowledge representation, perception, speech recognition, decision-making, language translation, motion, manipulation, intelligence, and creativity. The ability of machines to process natural language, to learn, to plan makes it possible for new tasks to be performed by intelligent systems. The main purpose of AI is to mimic the cognitive function of human beings and perform activities that would typically be performed by a human being. Without being taught by humans, machines use their own experience to solve a problem [1].

AI is a stand-alone independent electronic entity that functions much like a human expert. Today, AI is integrated into our daily lives in several forms, such as personal assistants, automated mass transportation, aviation, computer gaming, facial recognition at passport control, voice recognition on virtual assistants, driverless cars, companion robots, etc.

This chapter provides a comprehensive introduction to artificial intelligence. It begins by providing some historical background on AI. It then presents various components of AI. It briefly covers some common applications of AI. It discusses artificial general intelligence (AGI), weak AI, and strong AI. It also addresses the benefits and challenges of AI and AGI. The last section concludes with comments.

1.2   HISTORICAL BACKGROUND

From ancient times, humans have been dreaming of creating artificial intelligence. Artificial Intelligence somewhat scares and intrigues us. Early advocates of AI envisioned machines that had a wide variety of human capabilities. Modern AI research started in the mid-1950s when AI researchers were convinced that & machines will be capable, within twenty years of doing any work a man can do. In the 1970s, it was obvious that researchers had grossly underestimated the difficulty of the project. In 20 years, AI researchers have been shown to be fundamentally mistaken. By the 1990s, AI researchers expect that today's artificial intelligence will eventually evolve into artificial general intelligence (AGI) [1].

In the modern age, important events and milestones in the evolution of AI include the following [2]:

1950:   Alan Turing publishes Computing Machinery and Intelligence. In the paper, Turing—famous for breaking the Nazi's ENIGMA code during WWII—proposes to answer the question "can machines think?

1956: John McCarthy coins the term "artificial intelligence" at the first-ever AI conference at Dartmouth College. This is when the field of AI officially started.

1967: Frank Rosenblatt builds the Mark 1 Perceptron, the first computer based on a neural network that "learned" through trial and error. Just a year later, Marvin Minsky and Seymour Papert publish a book titled Perceptron, which becomes the landmark work on both neural networks and AI.

1980s: Neural networks featuring backpropagation — algorithms for training the network—become widely used in AI applications.

1997: IBM's Deep Blue beats then world chess champion Garry Kasparov, in a chess match.

2011: IBM's Watson captivated the public when it defeated two former Champions Ken Jennings and Brad Rutter on the game show Jeopardy!

2015: Baidu's Minwa supercomputer uses a special kind of deep neural network called a convolutional neural network to identify and categorize images with a higher rate of accuracy than the average human.

2016: DeepMind's AlphaGo program, powered by a deep neural network, beats Lee Sodol, the world champion Go player, in a five-game match. The victory is significant given the huge number of possible moves as the game progresses.

The latest focus on AI has given birth to natural language processing, computer vision, robotics, machine learning, deep learning, and more.

1.3 COMPONENTS OF ARTIFICIAL INTELLIGENCE

Today, AI is already everywhere, and it drives many aspects of our lives. AI is not a single technology but a range of computational models and algorithms. Artificial intelligence has the following main components [3,4]:

1.3.1 Expert systems:

Expert system (ES) was the first successful implementation of artificial intelligence and may be regarded as a branch of AI mainly concerned with a specialized knowledge-intensive domains like medicine. An expert system is computer software that simulates the judgment and behavior of a human expert. It is also known as an intelligent system or knowledge-based system. It encapsulates specialist knowledge of a particular domain of expertise and can make intelligent decisions. It has a knowledge base and a set of rules that infer new facts from the knowledge. Expert systems solve problems with an inference engine that draws from a knowledge base equipped with information about a specialized domain, mainly in the form of if-then rules. It is based on expert knowledge in order to emulate human expertise in any specific field. The basic concept behind ES is that expertise (such as a highly-skilled medical doctor or lawyer) is transferred from a human expert to a computer system. Non-expert users, seeking advice in the field, question the system to get expert knowledge [5].

Expert systems are widely used in industries. From 1982 - 1990, the Japanese government heavily funded expert systems and other AI-related endeavors as a part of their Fifth Generation Computer Project (FGCP). Expert systems are finding a wide range of applications due to their capability to provide solutions to a variety of real-life problems. They are widely used in healthcare, business, and manufacturing.

1.3.2   Fuzzy logic:

This makes it possible to create rules for how machines respond to inputs that account for a continuum of possible conditions, rather than straightforward binary. Where each variable is either true or false (yes or no), the system needs absolute answers. However, these are not always available. Fuzzy logic allows variables to have a" truth value" between 0 and 1. It uses approximate human reasoning in knowledge-based systems. It was introduced in 1960s by Lotfi Zadeh of University of California, Berkeley known as father of fuzzy set theory. Fuzzy logic is useful in manufacturing processes as it can handle situations that cannot be adequately handled by traditional true/false logic [6].

1.3.3
Neural networks

These are specific types of machine learning systems that consist of artificial synapses designed to imitate the structure and function of brains. An artificial neural network (ANN) is an information processing device that is inspired by the way the brain processes information. They were originally developed to mimic the learning process of the human brain. The idea of ANNs was inspired by the structure of the human brain and what the brain can do. They may be regarded as a sort of parallel processor designed to imitate the way the brain accomplishes tasks. They are made up of artificial neurons, take in multiple inputs, and produce a single output. The network observes and learns as the synapses transmit data to one another, processing information as it passes through multiple layers [7]. As shown in Figure 1.1, artificial neural networks are multi-layer fully connected neural nets.

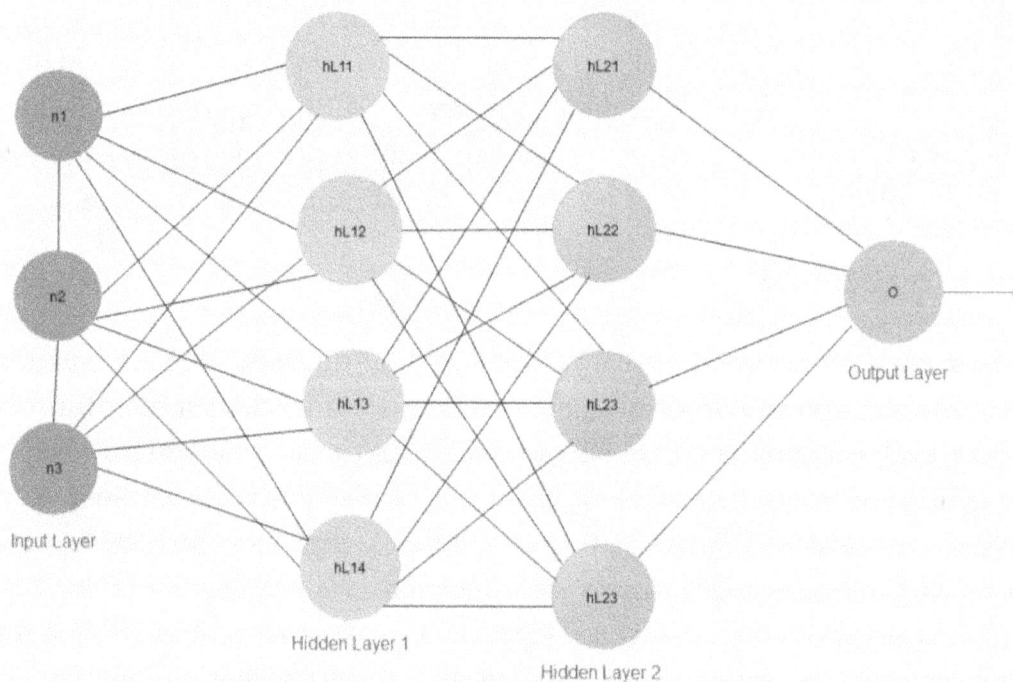

Figure 1.1 Artificial neural networks

Different types of artificial neural network are available: (1) support vector machine (SVM), (2) self-organization map (SOM), (3) multilayer perceptron (MLP). Typically, Neurons are organized in layers. Due to the fact ANNs can reproduce and model nonlinear processes, they have found several applications in a wide range of disciplines including system identification and control, quantum chemistry, pattern recognition, medical diagnosis, finance, data mining, machine translation, neurology, and psychology [8].

1.3.4 Machine Learning:

Machine learning (ML) is essentially the study of computer algorithms that improve automatically through experience. It is the field that focuses on how computers learn from data. This includes a broad range of algorithms and statistical models that make it possible for systems to find patterns, draw inferences, and learn to perform tasks without specific instructions. Machine learning is a process that involves the application of AI to automatically perform a specific task without explicitly programming it. Learning algorithms work on the assumption that strategies, algorithms, and inferences that worked well in the past are likely to work well in the future. ML techniques may result in data insights that increase production efficiency. Using ML can save time for practitioners and provide unbiased, repeatable results. Today, artificial intelligence is narrow and mainly based on machine learning. There are two types of learning: supervised learning and unsupervised learning. Supervised learning focuses on classification and prediction. It involves building a statistical model for predicting or estimating an outcome based on one or more inputs. It is often used to estimate risk. Supervised ML is where algorithms are given training data. Learning from data is used when there is no theoretical or prior knowledge solution, but data is available to construct an empirical solution. Supervised ML is increasingly being used in medicine such as in cardiac electrophysiology. In unsupervised learning, we are interested in finding naturally occurring patterns within the data. Unlike supervised learning, there is no predicted outcome. Unsupervised learning looks for internal structure in the data. Unsupervised learning algorithms are common in neural network models. Machine learning techniques have been currently applied in the analysis of data in various fields including medicine, finance, business, education, advertising, cyber security, and energy applications [9].

1.3.5 Deep Learning:

This is a form of machine learning based on artificial neural networks. Deep learning (DL) architectures are able to process hierarchies of increasingly abstract features, making them especially useful for purposes like speech and image recognition and natural language processing. Deep learning networks can deal with complex non-linear problems. It extracts complex features from high-dimensional data and applies them to develop a model that relates inputs to outputs. The most common form of deep learning architectures is multi-layer neural networks. Deep learning has many advantages over shallow learning. Due to this, deep learning networks have received much attention as they can deal with more complex non-linear problems. Recently, companies such as IBM, Microsoft, Google, Apple, and Baidu have invested in and developed deep learning. They have taken advantage of their massive data and large computational power to deploy deep learning on a large scale. Although DL has achieved some success and found applications in various fields, it is still in its infancy [10].

1.3.6 Natural Language Processors:

For AI to be useful to us humans, it needs to be able to communicate with us in our language. Computer programs can translate or interpret language as it is spoken by normal people. Language is crucial around the world in communication, entertainment, media, culture, drama, movie, and the economy. Natural language processing (NLP) refers to the field of study that focuses on the interactions between human language and computers. It is a computational approach to text analysis. It involves the study of mathematical and computational modeling of various aspects of language. It is an Interdisciplinary field involving computer science, linguistics, logic, and psychology.

NLP is important because of the major role language such as English plays in human intelligence and because of the wealth of potential applications. NLP is commonly used for text mining, machine translation, and automated question-answering. Applications of NPL include interfaces to expert systems and database query systems, machine translation, text generation, story understanding, automatic speech recognition, and computer-aided instruction. It also has great potential in healthcare, mobile technology, cloud computing, virtual reality, election, social work, and social networking [11].

1.3.7 Robots:

AI is heavily used in robots, which are computer-based programmable machines that have physical manipulators and sensors. Sensors can monitor temperature, humidity, pressure, time, record data, and make critical decisions in some cases. Robots have moved from science fiction to your local hospital. In jobs with repetitive and monotonous functions they might even completely replace humans. Robotics and autonomous systems are regarded as the fourth industrial revolution.

Robotics is a branch of engineering and computer science that involves the conception, design, manufacture, and operation of robots. A robot functions as an intelligent machine, meaning that, it can be programmed to take actions or make choices based on input from sensors. It involves using electronics, computer science, artificial intelligence, mechatronics, and bioengineering. Robots are applied in many fields including agriculture, education, manufacturing, entertainment, medicine, industry, space exploration, undersea exploration, sex, power grid, agriculture, construction, meat processing, household, mining, aerospace, electronics, and automotive. For example, robot police with facial recognition technology have started to patrol the streets in China. Future robots will operate in highly networked environments where they will communicate with other systems such as industrial control systems and cloud services [12]. These AI tools are illustrated in Figure 1.2.

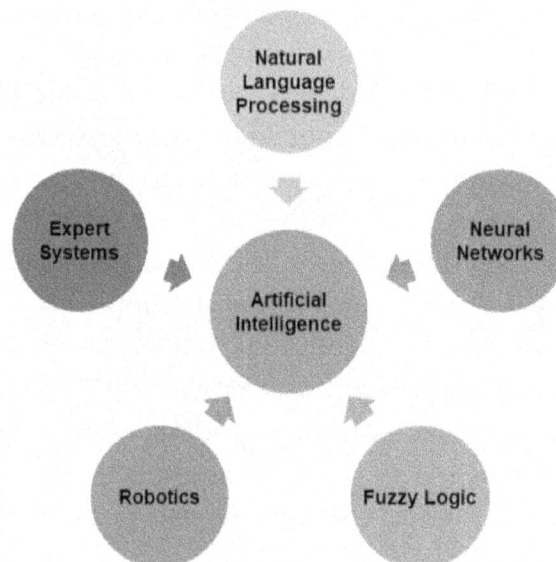

Figure 1.2 AI tools.

Each AI tool has its own advantages. Using a combination of these models, rather than a single model, is recommended. Other areas of AI research in evolutionary computation, artificial general intelligence, and explainable AI. Figure 1.3 show the relationship between AI and its tools.

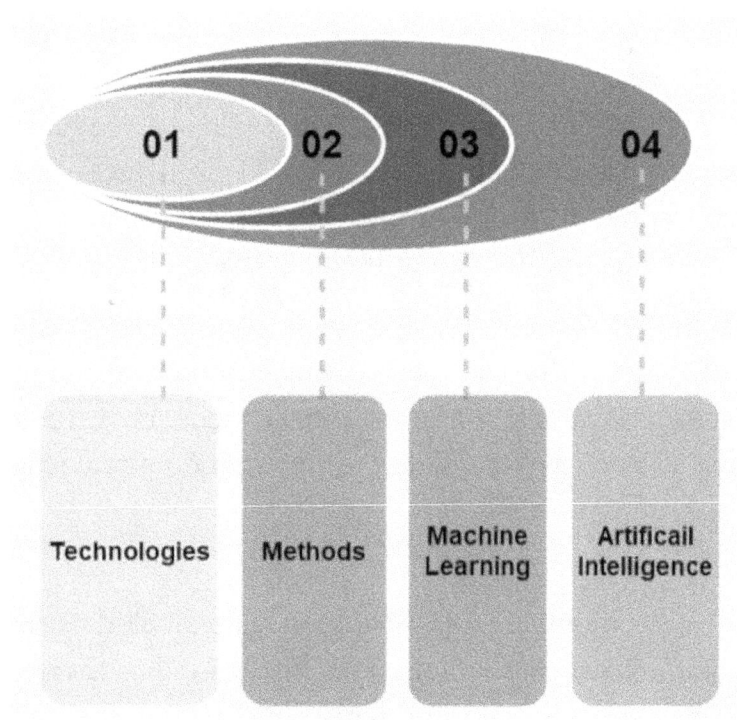

Figure 1.3 Relationship between AI and its tools.

1.4 APPLICATIONS

Artificial Intelligence (AI) is rapidly transforming our world. It is incorporated into a variety of different types of technology. AI has the potential to impact nearly all aspects of our society, including automation, healthcare, business, education, engineering, law, manufacturing, transportation, and security. These applications will be fully discussed in later chapters. Some of them are briefly discussed here [13]:

Automation: When paired with AI technologies, automation tools can expand the volume and types of tasks performed. The most common type was the automation of digital and physical tasks using robotic process automation technologies. Robots are intelligent machines used in automation to manufacture, assemble, paint, clean, and entertain.

Self-driving cars: Autonomous vehicles use a combination of computer vision, image recognition, and deep learning to build automated skills for piloting a vehicle. They are equipped with LIDARS (light detection and ranging) and remote sensors that gather information from the vehicle's surroundings. We can also expect to see driverless cars on the road in the next twenty years.

Autopilot Technology: This has been flying commercial and military aircraft for decades. Today, autopilot uses a combination of sensors, GPS technology, image recognition, collision avoidance technology, robotics, and natural language processing to guide an aircraft safely.

Healthcare: AI applications in healthcare are many, including using online virtual health assistants and chatbots to help patients and healthcare customers find medical information, schedule appointments, understand the billing process and complete other administrative processes.

Business: In the sphere of business, AI is poised to have a transformational impact. Chatbots have been incorporated into websites to provide immediate service to customers. Other programs, such as IBM Watson, have been applied to the process of buying a home. Banks are successfully employing chatbots to make their customers aware of services and offerings and to handle some transactions.

Government: AI in government is represented by intelligent software and hardware to boost smart government, reducing costs, minimizing corruption; and increasing transparency. Although AI will replace people from their jobs , it will create the need for other professionals as the data scientists for the government.

Education: AI technology in education is allowing a degree of flexibility and customization that was never before possible. It is revolutionizing schools and classrooms, making educator's jobs a lot easier. AI can automate grading and assess students. It is poised to revolutionize education.

Manufacturing: Manufacturing has been at the forefront of incorporating robots Into the workflow. Artificial intelligence appears to have significant application potential for many important manufacturing problems ranging from production planning and control processes to shop floor automation.

Transportation: AI is becoming a mega-trend in the transportation industry. AI technologies are used in transportation to manage traffic, predict flight delays, and make ocean shipping safer and more efficient.

Speech Recognition: This is AI technology that recognizes spoken words and converts them to digitized text.

Computer Vision: This is also known as machine vision. It helps transform otherwise, ordinary factories into intelligent systems and formidable competitive weapons. It emphasizes the development of techniques that allow a computer to recognize or otherwise understand the content of a picture. Computer vision provides the robot with its most flexible and powerful capability.

Other areas of application of AI include finance, ecommerce, social media, food industry, travel industry, engineering, law, gaming, agriculture, smart city, smart grid, security, and military.

1.5 ARTIFICIAL GENERAL INTELLIGENCE

The greatest fear about AI is producing a system capable of human-level thinking, known as artificial general intelligence (AGI), in which computers meet or even exceed human intelligence. AGI is a newly emerging field that aims at building "thinking machines" with intelligence comparable to that of humans.

The goal of AGI research is the development and demonstration of systems that exhibit the broad range of general intelligence found in humans. AGI does not exist right now [14].

An artificial general intelligence (AGI) refers to an intelligent machine capable of understanding the world as well as humans.

AGI has the capacity of an engineered system to display intelligence that is not tied to a highly specific set of tasks. These machines would be able to take over every task performed by humans. AGI is an AI that can at least match human intelligence's capabilities. Human intelligence can do the following: abstract reasoning, learning from experience, consciousness, composition of elements, adaptability to new environments, creativity, empathy, perception, problem-solving, navigation, communication, etc. Thus, some capabilities that will turn AI into AGI include [15]:

- Sensory Perception: While deep learning has enabled some advances in computer vision, AI systems are still far away from developing human-like sensory-perception capabilities.

- Human Perception: AI systems are not yet able to replicate this distinctly human perception.

- Fine Motor Skills: Very few humans would let any of the robot manipulators or humanoid hands do that task for us.

- Natural Language Understanding: Machine will interact with humans. If AI lacks understanding of natural language, it will not be able to operate in the real world.

Other necessary capabilities include problem-solving, creativity or originality, navigation, etc.

Some of the main issues addressed in AGI research include [16]:

- Value specification: How do we get an AGI to work towards the right goals?
- Reliability: How can we make an agent that keeps pursuing the goals we have designed it with?
- Corrigibility: If we get something wrong in the design or construction of an agent, will the agent cooperate with us trying to fix it?
- Security: How to design AGIs that are robust to adversaries and adversarial environments?
- Safe learning: AGIs should avoid making fatal mistakes during the learning phase.
- Intelligibility: How can we build agents whose decisions we can understand?
- Societal consequences: AGI will have substantial legal, economic, political, and military consequences. How should we manage the societal consequences?

AGI is an emerging technology that has the potential of massively benefiting the society. It can be used to solve complex problems. It could also be used as a cyber weapon, a means of monitoring, influencing, and controlling society. It could be used to handle economic crises, overthrow a nation, exploit, plan, and lead a military invasion [17]. The Artificial General Intelligence Society (http://www.agi-society.org/) is a nonprofit organization with the following goals:

- to promote the study of artificial general intelligence (AGI), and the design of AGI systems.
- to facilitate co-operation and communication among those interested in the study and pursuit of AGI
- to hold conferences and meetings for the communication of knowledge concerning AGI
- to produce publications regarding AGI research and development
- to publicize and disseminate by other means knowledge and views concerning AGI.

The future of AGI is a hard, complicated issue. No one can sensibly claim to know what is going to happen. It is safe to assume that AGI is likely to emerge gradually.

1.6 WEAK AND STRONG AI

Two different approaches have developed in the history of AI. In 1980, John Searle introduced the terms "weak AI" and strong AI. Narrow AI is what we have now, while general AI is what we wish to achieve. These are also referred to as specialized and general AI.

A strong artificial intelligence is in principle identical to human intelligence, i.e., strong AI can think and have a mind. Strong AI can be developed by combining the programs that solve various sub-problems. Strong AI is for machines capable of experiencing consciousness, which is the capacity to recall memories and dream about the future. Strong AI is now referred to as AGI. Some AGI projects are currently underway, including DeepMind, OpenCog, and OpenAI. The IEEE has developed its own recommendations for building safe AGI systems, which include that AGI systems should be transparent, that a safe and secure environment should be developed, and that such systems should resist being shut down by operators.

Weak AI is everywhere around us and is the most successful realization of AI to date. A weak artificial intelligence is less ambitious than strong AI, and therefore less controversial, i.e., weak, or narrow AI can (only) act like it thinks and has a mind. Weak or specialized AI is the application of AI to specific tasks such as industrial robots, virtual personal assistants, such as Apple's Siri, Internet searching, driving a car, or playing a video game. Weak AI is sometimes called artificial narrow intelligence (ANI). Weak AI is AI as known today. Weak AI is limited to the use of software to study or accomplish specific problem-solving. IBM's Watson supercomputer, expert systems, and the self-driving car are all examples of weak AI. Freely accessible weak AIs include Google AI or Apple's Siri and others. Today, narrow AI tools have become mainstream in business and society [18]. The progression of artificial intelligence (AI) is shown in Figure 1.4.

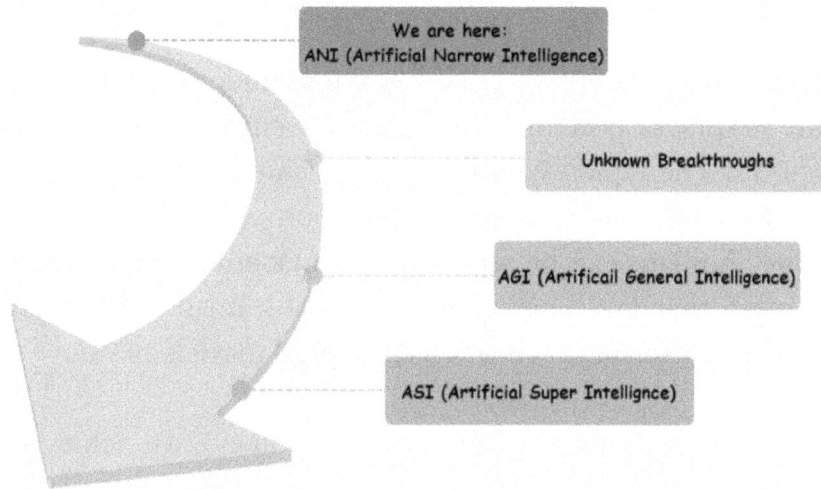

Figure 1.4 The progression of artificial intelligence (AI).

1.7 BENEFITS

Artificial intelligence enables machines to learn from experience, adjust to new inputs, and perform human-like tasks. It adds intelligence to existing products. It is reshaping economies, promising to boost productivity, improve efficiency, and lowest costs. It contributes to better lives and helps people make better predictions and more informed decisions. People will be capable of buying goods from anywhere in the world using the Internet and exploiting the benefits of availing AI technologies.

AGI will have the same general capabilities as a human being and augment the advantages that computers have over humans today -- the perfect recall, flexible thinking, reasoning with computational advantages, recognizing language, understanding speech, describing photos and videos, and split-second number crunching. AGI technology enables an AI agent to acquire various problem-solving skills through learning. AGI would ultimately render human labor obsolete [19]. It has been observed that AI systems are rapidly becoming super intelligent. "Superintelligence" refers to an intellect that is much smarter than the best human brains. AGI is a likely path to superhuman intelligence. For this reason, some advocates claim that AGI could be the most powerful technology ever invented.

1.8 CHALLENGES

From its inception, artificial intelligence has come under scrutiny from scientists and the public alike. In spite of potential risks, there are currently few regulations governing the use of AI tools. The rapid evolution of AI technologies is a barrier to developing regulation of AI. A popular contentious issue on AI is how it may affect human employment since humans can be replaced by a machine. The robot misconception is related to the myth that machines cannot control humans. While emotion sums up human experiences, there are no emotions in typical models of AI.

There are doubts about whether strong AI will ever be achieved. In spite of the advocates of near-future AGI, there are several challenges facing AGI. One downside of AGI is empowering surveillance and control of populations, underpinning fearsome weapons, and removing the need for governments to look after the obsolete populace. AGI is a fairly abstract and vague concept [20].

An AI arms race is difficult to be stopped, only managed. Since an AI arms race could compromise safety precautions during the research and development of AGI, an arms race could prove fatal to the entire human species. The race for an AGI may result in a poor-quality AGI that does not fully consider the welfare of humanity. These challenges have impeded the progress of AI and AGI. As of November 2020, AGI remains a speculative idea and is yet to be constructed. It is therefore worthwhile to investigate how to maximize the benefits while addressing the challenges [21,22].

1.9 CONCLUSION

Artificial intelligence refers to any human-like intelligence exhibited by a computer, robot, or another machine. After a long time, of perception of AI as science fiction, AI is now part of our everyday lives. Artificial intelligence has the potential to change everything. AI has attracted attention as a key for growth and development in developed nations such as Europe and the United States and developing nations such as China and India.

Artificial general intelligence (AGI) is a newly emerging field that aims at building "thinking machines" with intelligence comparable to that of humans. It is essentially a Hypothesized system that could replicate any task now requiring human intelligence. Although advocates argue that they will be able to realize AGI using deep learning and big data, we have not come much closer to developing AGI. The future of artificial intelligence technology has broad application prospects, and it covers almost all areas of endeavor.

Some academic institutions are now offering courses on AI. More information on AI is available in the books in [23-45] and the following related journals:

- Artificial Intelligence
- Journal of Artificial Intelligence and Consciousness
- AI Magazine
- AI & Society
- Artificial Intelligence in Agriculture
- IEEE Transactions on Artificial Intelligence
- Journal of Artificial General Intelligence
- Journal of Experimental & Theoretical Artificial Intelligence

REFERENCES

[1] **M. N. O. Sadiku**, "Artificial intelligence", IEEE Potentials, May 1989, pp. 35-39.

[2] IBM Cloud Education, "Artificial intelligence (AI)," June 2020, https://www.ibm.com/cloud/learn/what-is-artificial-intelligence

[3] **M. N. O. Sadiku**, **Y. Zhou**, and **S. M. Musa**, "Natural language processing in healthcare," International Journal of Advanced Research in Computer Science and Software Engineering, vol. 8, no. 5, May 2018, pp. 39-42.

[4] "Applications of AI and machine learning in electrical and computer engineering," July 2020, https://online.egr.msu.edu/articles/ai-machine-learning-electrical-computer-engineering applications/#:~:text=Machine%20learning%20and%20electrical%20engineering,can %2 0%E2%80%9Csee%E2%80%9D%20the%20environment.

[5] **M. N. O. Sadiku**, **Y. Wang**, **S. Cui**, **S. M. Musa**, "Expert systems: A primer," International Journal of Advanced Research in Computer Science and Software Engineering, vol. 8, no. 6, June 2018, pp. 59-62.

[6] **G. Singh**, **A. Mishra**, and **D. Sagar**, "An overview of artificial intelligence," SBIT Journal of Sciences and Technology, vol.2, no. 1, 2003.

[7] **A. Dertat**, "Applied deep learning - Part 1: Artificial neural networks," August 2017, https://towardsdatascience.com/applied-deep-learning-part-1-artificial-neural-networks-d7834f67a4f6

[8] **M. N. O. Sadiku**, **S. M. Musa**, and **O. S. Musa**," Neural Networks in the Chemical Industry," Invention Journal of Research Technology in Engineering and Management, vol. 1, no. 12, version 3, Nov. 2017, pp. 25-27.

[9] **M. N. O. Sadiku**, **S. M. Musa**, and **O. S. Musa**, "Machine learning," International Research Journal of Advanced Engineering and Science, vol. 2, no. 4, 2017, pp. 79-81.

[10] **M. N.O. Sadiku**, **M. Tembely**, and **S.M. Musa**," Deep learning," International Research Journal of Advanced Engineering and Science, vol. 2, no. 1, 2017, pp. 77,78.

[11] **M. N. O. Sadiku**, **Y. Zhou**, and **S. M. Musa**, "Natural language processing," International Journal of Advances in Scientific Research and Engineering, vol. 4, no. 5, May 2018, pp. 68-70.

[12] **M. N. O. Sadiku**, **S. Alam**, and **S.M. Musa**, "Intelligent robotics and applications," International Journal of Trends in Research and Development, vol. 5. No. 1, January- February 2018, pp. 101-103.

[13] "Artificial intelligence," https://searchenterpriseai.techtarget.com/definition/AI-Artificial-Intelligence

[14] **M. N. O. Sadiku**, O. **D. Olaleye**, **A. Ajayi-Majebi**, and **S. M. Musa**, Artificial General Intelligence: A primer, & International Journal of Trend in Research and Development, vol. 7, no. 6, Nov.-Dec. 2020, pp. 7-9.

[15] **F. Berruti**, **P. Nel**, and R. **Whiteman**, "An executive primer on artificial general intelligence," April 2020, https://www.mckinsey.com/business-functions/operations/our-insights/an-executive-primer-on-artificial-general intelligence

[16] **A. Katte**, "Can morality be engineered in artificial general intelligence systems?" October 2018, https://analyticsindiamag.com/can-morality-be-engineered-in-artificial-general- intelligence-systems/

[17] **M. Bullock**, "Artificial general intelligence in plain English," October 2019, https://towardsdatascience.com/artificial-general-intelligence-in-plain-english-e8f6e9a56555

[18] **T. Angrignon**, "AI definitions, and narrow vs. general vs. super intelligence," June 2020, https://productiveai.com/ai-definitions-and-narrow-vs-general-vs-super-intelligence/

[19] **N. Heath**, "What is artificial general intelligence?" August 2018, https://www.zdnet.com/article/what-is-artificial-general-intelligence/

[20] **W. Pei**, "Artificial general intelligence——A gentle introduction," 2007 https://www.semanticscholar.org/paper/Artificial-General Intelligence%E2%80%94%E2%80%94a-Gentle Pei/b8970f54a8795df1670d633150be34bc4b35b9eb https://www.zdnet.com/article/what-is-artificial-general-intelligence/

[21] **R. Fjelland**, "Why general artificial intelligence will not be realized," Humanities and Social Sciences Communications, vol 7, no, 10, 2020.

[22] **P. Torres**, "The possibility and risks of artificial general intelligence," Bulletin of the Atomic Scientists, vol. 75, no. 3, 2019, pp. 105-108.

[23] **S. Marsland**, Machine Learning: An Algorithmic Perspective. Boca Raton, FL: CRC Press, 2nd edition, 2018.

[24] **K. Hammond**, Practical Artificial for Dummies, Narrative Science Edition Hoboken, NJ: John Wiley & Sons, 2015.

[25] **M. Hutter**, Universal Artificial Intelligence: Sequential Decisions Based on Algorithmic Probability. Springer, 2005.

[26] **B. Goertzel** and **P. Wang** (eds.), Advances in Artificial General Intelligence: Concepts, Architectures and Algorithms. IOS Press, 2007.

[27] **E. R. Ranschaert, S. Morozov**, and **P. R. Algra** (eds.), Artificial Intelligence in Medical Imaging Opportunities, Applications and Risks. Springer 2019

[28] **R. Mitchell, J. Michalski**, and **T. Carbonell**, An artificial Intelligence Approach Springer, 2013.

[29] **P. Joshi,** Artificial Intelligence with Python: A Comprehensive Guide to Building Intelligent Apps for Python Beginners and Developers Paperback. Packt Publishing, 2017.

[30] **V. Dignum**, Responsible Artificial Intelligence: How to Develop and Use AI in a Responsible Way. Springer 2019.

[31] **J. Finlay** and **A. Dix**, An Introduction to Artificial Intelligence. Boca Raton, FL:CRC Press, 2020.

[32] **K Warwick,** Artificial Intelligence: The Basics. Routledge, 2013.

[33] **S. Russell** and **P. Norvig**, Artificial Intelligence: A Modern Approach. Pearson, 4th Edition, 2020.

[34] **M. Mitchell,** Artificial Intelligence: A Guide for Thinking Humans. Farrar, Straus and Giroux, 2019.

[35] **N. J. Nilsson**, Principles of Artificial Intelligence. Elsevier, 2014.

[36] **N. J. Nilsson**, The Quest For Artificial Intelligence. Cambridge University Press, 2009.

[37] **A. Barr** and **E.A. Feigenbaum**, The Handbook of Artificial Intelligence: Volume 2. Elsevier, 2014.

[38] **J Copeland**, Artificial intelligence: A Philosophical Introduction. Blackwell Publishers, 2015.

[39] **W Ertel**, Introduction to Artificial Intelligence. Springer, 2018

[40] **J Haugeland**, Artificial intelligence: The Very Idea. MIT Press, 1989.

[41] **M Ginsberg**, Essentials of Artificial Intelligence. Elsevier, 2012.

[42] **M. Flasiński**, Introduction To Artificial Intelligence. Springer, 2016.

[43] **D. Li** and **Y. Du**, Artificial Intelligence With Uncertainty. Boca Raton, FL: CRC Press, 2nd edition, 2017.

[44] **B. Goertzel** and **C. Pennachin**, Artificial General Intelligence. Springer, 2007.

[45] B. **Goertzel**, AGI Revolution: An Inside View of the Rise of Artificial General Intelligence. Humanity Press, 2016.

CHAPTER 2
AI IN SMART CITIES

"Smart cities are those who manage their resources efficiently. Traffic, public services, and disaster response should be operated intelligently in order to minimize costs, reduce carbon emissions, and increase performance." - *Eduardo Paes*

2.1 INTRODUCTION

Cities are centers for economic growth, job creation, new ideas incubator, technological evolution, networking, information sharing, and social transformation. By 2008, half of the world's population lived in cities. The continued growth of urbanization presents new challenges. Higher urbanization rates with 10 million inhabitants or more make it difficult to create a sustainable and cost-effective environment and a high quality of life for the citizens. Many cities of the world have infrastructure that is not adequate for meeting the needs of the growing urban population. Figure 2.1 shows a typical traditional city [1], while Figure 2.2 displays a modern city [2].

Figure 2.1 A typical traditional city [1].

Figure 2.2 A typical modern city [2].

In recent years, cities have undergone various changes due to the emergence and adoption of numerous concepts such as sustainable cities, inclusive cities, resilient cities, etc. Today, the focus is on smart cities, which aspire to connect all aspects of urban life. Smart city is a concept that leverages technology and intelligent insights from sensors to make life more comfortable and secure for residents. It reduces the urban consumption of resources, prevents wastage, and enhances comfort and security for residents.

The concept of smart cities provides a potential solution to the challenges. Many municipalities worldwide are turning to the smart cities concept to improve their infrastructure, communication, and services for residents. The rapid development of new technologies such as AI, big data, cloud computing, blockchain, 5G, ICT, and IoT is helping to drive the evolution of smart cities [3]. While several technologies are needed to realize the concept of smart cities, the two key enablers of the smart city are the Internet of things (IoT) and artificial intelligence (AI). Figure 2.3 shows the various components of a smart city. Prioritizing changes to urban infrastructure and investing in reliable technology is central to smart city development [4].

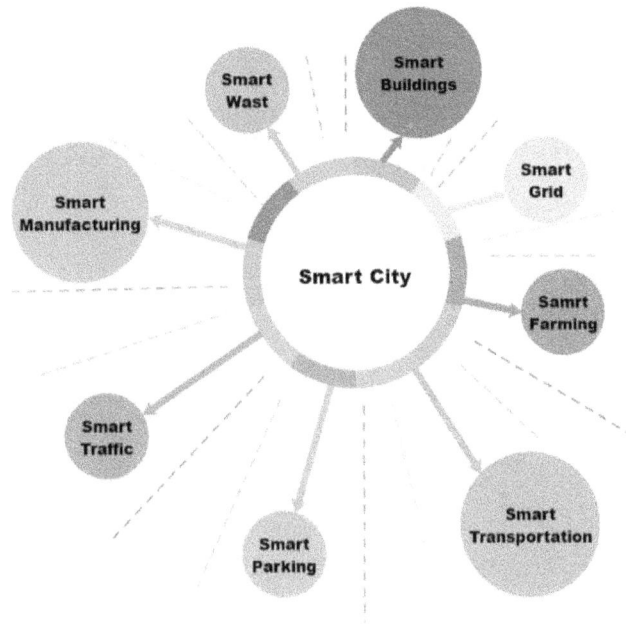

Figure 2.3 Components of a smart city [3].

This chapter examines the role of AI in the different sectors of the smart cities. It begins by giving a brief review of AI, particularly in relation to cities. It covers AI in smart city and the enabling technologies for AI used in smart cities. It then addresses different ways AI is applied in smart cities. It gives some examples on AI in smart cities. It highlights the benefits and challenges of AI in smart cities. It covers the future of AI in smart cities. The last section concludes with comments.

2.2 REVIEW ON ARTIFICIAL INTELLIGENCE

The term "artificial intelligence" (AI) was first used at a Dartmouth College conference in 1956. AI is now one of the most important global issues of the 21st century. AI is the branch of computer science that deals with designing intelligent computer systems that mimic human intelligence, e.g. visual perception, speech recognition, decision-making, and language translation [5]. As shown in Figure 2.4, AI is a multi-disciplinary field.

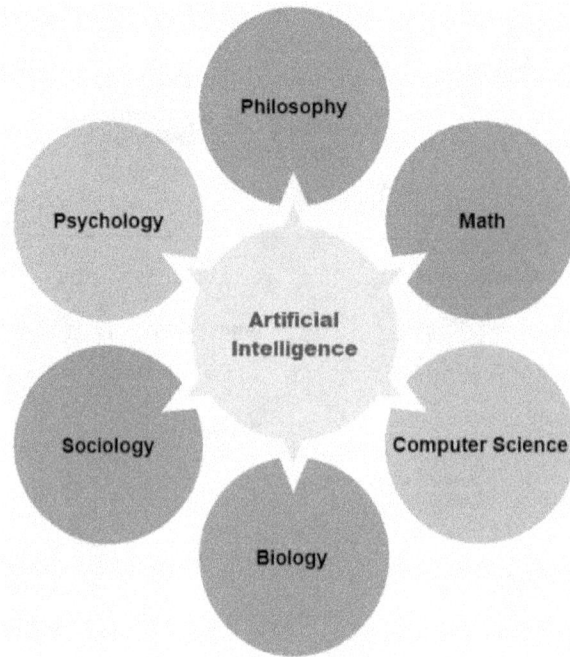

Figure 2.4 Artificial intelligence (AI) is a multi-disciplinary field.

The ability of machines to process natural language, to learn, to plan makes it possible for new tasks to be performed by intelligent systems. The main purpose of AI is to mimic the cognitive function of human beings and perform activities that would typically be performed by a human being.

AI is a collection of programmed algorithms to mimic human decision making. The components of AI include:

- Expert systems
- Fuzzy logic
- Neural networks
- Machine learning

- Deep learning
- Natural Language Processors
- Robots

All AI systems rely on algorithms, which are basically a set of instructions on how to organize and manage data. AI is well-suited to form the analytical foundation of smart city programs [6].

2.3 AI IN SMART CITIES

Cities serve as the engine for human prosperity and the vital hubs to exchange goods, services, and even ideas. No city is the same. Different cities require different approaches to become smart. Therefore, the conceptualization and implementation of the smart city concept may vary from city to city and country to country depending on the level of development.

Everything is getting smart - automobiles, homes, industries, health care, agriculture, and cities. A smart city is an urban area that uses intelligent systems to ease up the livelihood of its people. It relies on information and communication technologies to build economic growth and improve quality of life. It consists of safe, secure, and reliable interconnected sensors, actuators, and relays which collect, process, and transmit data. Cities have wealth of data sources, such as ticket sales on mass transit, local tax information, police reports, sensors on roads and local weather stations. The raw data generated is unimaginably bigger than what humans can analyze and process. This is where the role of AI comes in. AI is being used in the design and infrastructure management of cities.

The Smart Cities Council promotes three core values that make a city smart [7]:

- Livability: Cities that provide clean, healthy living conditions without pollution and congestion. This is achieved with a digital infrastructure that makes city services instantly and conveniently available anytime, anywhere. The livability aspect of cities is influenced by the security and safety therein. Culture, governance, and metabolism form the basis of livable smart cities.
- Workability: Cities that provide the enabling infrastructure — energy, connectivity, computing, essential services — to compete globally for high-quality jobs.
- Sustainability: Cities that provide services without stealing from future generations. A smart city ensures economic growth and quality of life.

2.4 SMART CITY TECHNOLOGIES

Currently, many cities are being developed under the smart city platform. Smart cities are complex systems designed by using highly advanced integrated technologies which include millions of sensors and devices. They make use of various technologies including artificial intelligence, Internet of things (IoT), big data, and Clouds of things (CoT), and Information and Communication Technologies (ICT) [8-10].

- **Artificial Intelligence (AI):** AI offers the ability to sort through and analyze big data using algorithms to solve problems, predict or come to logical conclusions. AI can keep a count of any number of vehicles, pedestrians or any other movements while keeping a track on their speeds. It can carry out face recognition, read license plates, and process satellite data. Smart city uses AI to optimize city functions, drive economic growth, improve quality of life, and underpin governance structures.

- **Internet of Things (IoT):** The IoT is considered as the next big step in the evolution of the Internet. The IoT technology is expected to integrate the Internet into a multitude of things (e.g., objects, environments, vehicles, and clothing). The combination of AI and IoT has the potential to address key challenges posed by an excessive urban population. With IoT, AI will become the standard way we design most things.

- **Big Data:** Data generation is believed to be possible in almost all sectors of human activity. The use of IoT and other technologies in smart cities generates a huge amount of data. This big data needs to be properly analyzed and managed to extract patterns, which are useable for various services such as public health, public information systems, city management, energy efficiency, transport, security and emergency services, waste management, and water management. Big data can be useful in optimal usage of resources while making informed decisions.

- **Information and Communication Technologies (ICTs):**

Smart cities need ICT technologies as a core to be able to handle the innovative smart city challenges. ICTs are used to stimulate economic growth and improving the quality of life. They can also be used to integrate all hardware and software technologies to improve urban management. A valuable smart city ICT infrastructure must be able to integrate the smart homes into a coherent smart city concept. An ICT- based infrastructure must comprise technologies such as IoT, CoT and distributed AI. Figure 2.5 illustrates some of these technologies.

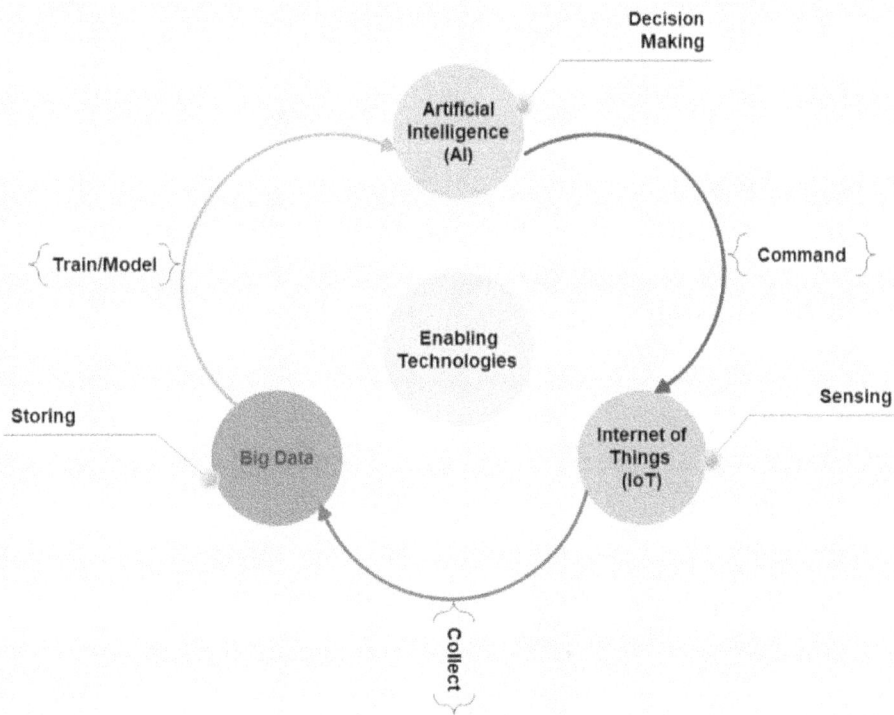

Figure 2.5 Some enabling technologies.

Any city can be transformed into a smart city with the help of these technologies.

2.5 APPLICATIONS OF AI IN SMART CITIES

A city is made "smart" when it uses intelligent or smart technology to improve infrastructure and services in a city. A technology that is helping accelerate the realization of the smart city concept is artificial intelligence (AI). These are some of the following ways AI is helping to realize the smart city concept [11-14].

•   **Smart Grid:** Cities are the major consumers of electricity today. The smart grid is an electrical power grid that is integrated with an AI-enabled, two-way communication network providing energy and information. AI is a powerful tool to push smart grid into the new generation of power systems. It provides a platform for clean, sustainable, efficient, and reliable energy generation, delivery, and consumption. Smart power grid and smart water management helps to produce energy with less pollution in smart cities.

•   **Smart Governance:** The smart city infrastructure is not complete without smart governance. Smart governance utilizes ICT intelligently along with data, evidence, and other resources to improve decision making and compliance towards the needs of the citizens. Smart governance increases in complexity due to the evolving dynamics of policy.

•   **Smart Parking:** In cities, parking solutions optimize parking spaces in a way that is efficient, smart, and contributes to a better traffic flow. AI can improve parking management in cities by providing authorities with data regarding the availability or unavailability of parking spots. The technology helps authorities to make parking more efficient by allowing them to identify patterns of use of parking spaces. A smart parking app can help the driver find available parking spaces with ease. VIMOC is using AI to make parking easier in Redwood City.

•   **Smart Home:** A smart home is a residence equipped with technologies which include development of smart home technologies contributed to the transition of the home from traditional to a smart internet-connected one. Smart homes can be designed to monitor and control home activities for convenience, provide occupants with better comfort, and possibly reduce energy use. AI technology is used in home products to make them smart.

- **Public Transit**: Urban AI can help harmonize the experience of its riders. Passengers of trains, buses, and cars can provide real-time information through their mobile apps to communicate delays, breakdowns, and less congested routes. Mobility data can translate into improved walkability and reduced commute times and make it a win-win-win for all traffic participants. For example, Dubai developed a number of Smart City projects, one of which monitored the condition of bus drivers.

- **Security Surveillance:** The safety and security of the citizens are at topmost priority in cities. AI enables real-time monitoring and decision-making, which can help public safety in cities. AI-enabled cameras and sensors can keep an eye on the surroundings to enhance the safety and security. Autonomous flying drones with in-built cameras can track humans, monitor traffic movement and provide the 2D aerial view imagery mapping. The technology is used by law enforcement agencies for crime prevention. Police departments around the world are using AI Technologies to undertake policing functions in an efficient manner.

- **Traffic Congestion:** Traffic congestion is rising in cities worldwide and is a major challenge in many cities. To solve this problem, cities are turning to the use of IoT and AI-enable traffic solutions. AI-supported traffic sensor systems can use cameras to collect real-time data of vehicles on road. Uber is now using AI to give a better riding experience to its customers.

- **Waste Management:** People living in cities generate a lot of waste. There is a need to manage garbage, keep the environment clean, and maintain sound hygiene in cities. By combining AI and IoT, it becomes easier for city authorities to remotely monitor waste levels.

- **Autonomous Vehicles:** Perhaps the most revolutionary AI application in transport today is the automation of vehicles. Here the artifact where the artificial intelligence resides is a car or vehicle. Autonomous vehicles are equipped with sensors which make them capable of perceiving the built environment. The AI-enabled vehicle can sense the surrounding urban environment by means of cameras, radars, and lidar systems. The artificial intelligence employs downloadable data to drive the vehicle to a given location, with no human input or supervision. The transition to autonomous cars may likely reshape urban mobility. Figure 2.6 shows a typical autonomous vehicle [15].

Figure 2.6 A typical autonomous vehicle [16].

Other areas of application of AI in smart cities include urban governance, smart policing, AI-powered robots, drones, healthcare, smart health, smart tourism, smart building, home automation, cybersecurity, combating social problems such as homelessness, ridesharing, personal assistant, and crowdsourced data collection.

2.6 EXAMPLES OF AI SMART CITIES

Smart cities, with rapid increase in urban growth, is a major challenge for developed and developing nations. Examples of AI use in smart cities include [16]:

•   Amsterdam is now mounting smartphones on waste collection vehicles and citizens' bikes to capture and stream images of the conditions on the streets to a server, which can use AI to detect abnormalities.

•   Montreal is mounting cameras and sensors on public vehicles, such as buses, fire engines, and snow ploughs, to collect data that can be used to improve urban services.

• Singapore incorporates smart policing in their city initiative by installing a network of cameras and sensors in almost every corner. This helps to identify people who are smoking in prohibited zones.

• In India, ridesharing companies like Uber and Ola permit constant recording and monitoring of every cab, thus offering safety. In the event of any crime, police get instantaneous alerts.

• In Egypt, the government is building a brand-new capital outside of Cairo that will soon become the smartest city in Africa. That system will rely on ubiquitous sensors for continuous monitoring and management of traffic, utilities, public safety, and more.

2.7 BENEFITS

The application of AI in smart cities has several benefits for humans as well as to the environment. AI benefits a city's transportation, education, environmental protection, public safety, and more. Combining AI with IoT can provide intelligent systems that cities need to minimize straining existing infrastructure, prevent deterioration of the environment, minimize traffic congestion, reduce urban crime, improve sanitation, reduce pollution, ease parking management, and enhance comfort and security for residents. Some benefits can come in many forms, including reduced crime, cleaner air, more orderly traffic flow, and more efficient government services.

AI is playing a big role in making urbanization smarter by making cities equipped with advanced features for citizens to live, walk, shop, and enjoy a safe environment. Smart city transition not only creates jobs, but also helps save the environment, reduce energy expenditure, and generate more revenue. Other benefits include [17].

• AI will curb pollution and reduce CO_2 emission.

• Many cities are turning to smart technologies to ease traffic congestion and optimizing mobility.

• AI in security cameras or drones can recognize the human faces and trace their identity.

• AI technologies are taking the management of urban services out of the hands of humans, operating the city in an autonomous way.

• AI creates more sustainable cities, maintain infrastructure, and improve public services for residents and visitors

2.8 CHALLENGES

While cities provide certain opportunities for the dwellers, there are also costs associated with fitting so many people into the same place. Achieving the AI in smart cities is an uphill task with several challenges like economic inequality, administration, sanitation, traffic congestion, security surveillance, parking management, and data privacy risks. The challenges of running a city are getting tougher as cities continue to grow. Other challenges include [18]:

• AI as a technology has limits and its applicability has been slow in adoption.

• The devices and equipment, and sensor technology required to develop smart cities are sophisticated and costly.

• Monitoring people or watching their activities can be a challenge as the use of IoT and sensor technology increases.

• The threat of cyber-attacks is a critical issue for smart cities. Awareness, education, and transparency on the purpose of data collection are crucial.

• Covering all age, gender, class and income group of people from society is necessary.

• Making urban AI a reality involves several challenges in the areas of privacy, security, accuracy, and bandwidth.

• Civic leaders must address economic inequality head-on.

It has been argued that the more revolutionary and disruptive a smart technology is, the more likely will be the transformation of the city that integrates it.

2.9 THE FUTURE OF AI IN SMART CITIES

Individuals, city officials, policymakers, and futurists have long envisioned smart cities of the future where residents and visitors thrive. Smart cities are designed to efficiently manage growing urbanization, energy consumption, maintain a green environment, improve the economic and living standards of their citizens. Cities should plan for a smooth transition toward the future in two ways.

First, they should have a short-term of investing in the technological innovations that are already available on the market and integrate them into the existing systems. Second, cities should have a long-term plan characterized by infrastructure and vehicles communicating with each other and with people [15]. For example, cities may update their fiber infrastructure for the purpose of better traffic management and for future connected and automated vehicles.

The future of smart cities relies heavily on artificial intelligence. AI and IoT are becoming increasingly integral to how the world operates. Much advancement in AI within the smart city will come from the automation of the routine. In the future of smart cities, everything from parking, waste management, traffic management, public safety, water and energy management, and citizen engagement is "smart" and efficient.

Technological advances are constantly reshaping the realization of smart-city initiatives which are being implemented across different geographical locations. Various technologies such as information and communication technology (ICT), artificial intelligence (AI), machine learning (ML), robotics, and the Internet of things (IoT) will play an outstanding role in realizing the concept of a smart city. While some progress has been made toward the future, cities continue to face complex challenges, including infrastructure maintenance, population growth, and environmental degradation.

2.10 CONCLUSION

A smart city is one that makes use of information and technologies to enhance the quality and performance of urban services. The transformation from traditional to smart cities will be spearheaded by AI applications. AI technology is accelerating far more quickly than anyone could have imagined. Organizations that have implemented AI technology is reaping its consistent benefits. City planners and engineers are now using AI and IoT to deal with increasingly complex environments. They should use AI tools for the welfare of citizens. For more information about artificial intelligence in cities, one should consult the books in [19-22] and related journals:

- Smart Cities
- Frontiers in Sustainable Cities
- City Pulse

REFERENCES

[1] **E. C. Manasseh**, "Combined artificial intelligence and IoT for smart sustainable cities,"https://www.itu.int/en/ITU-T/Workshops-and-Seminars/gsw/201804/Documents/Manasseh_Presentations.pdf

[2] **B. Chan**, "Ten best practices for building smart city innovation labs," April 11, 2019, https://www.iotcentral.io/blog/list/tag/smart+cities

[3] "Artificial intelligence for smart cities," August 2010, https://becominghuman.ai/artificial-intelligence-for-smart-cities-64e6774808f8

[4] **M. N. O. Sadiku, A. E. Shadare, E. Dada**, and **S. M. Musa**, "Smart cities," International Journal of Scientific Engineering and Applied Science, vol. 2, no. 10, Oct. 2016, pp. 41-44.

[5] **Y. Wang**, "Interdisciplinary study for future artificial intelligence development," Feb 16, 2018 https://medium.com/anth374s18/interdisciplinary-study-for-future-artificial-intelligence-development-28277e6ce9a5

[6] **A. Tomer**, "Artificial intelligence in America's digital city," July 2019, https://www.brookings.edu/research/artificial-intelligence-in-americas-digital-city/

[7] **T. V. Ark**, "How cities are getting smart using artificial intelligence," June 2018, https://www.forbes.com/sites/tomvanderark/2018/06/26/how-cities-are-getting-smart-using-artificial-intelligence/?sh=6b3ebc633803

[8] "Smart city, IoT and AI," https://riberasolutions.com/smart-city-iot-and-ai/

[9] **M. N. O. Sadiku, O. D. Olaleye, A. Ajayi-Majebi**, and **S. M. Musa**, "Artificial intelligence in smart cities & International Journal of Trend in Research and Development, vol. 8, no. 1, Jan.-Feb. 2021, pp. 6-9.

[10] **K. E. Skouby et al.**, "How IoT, AAI can contribute to smart home and smart cities services: The role of innovation," Proceedings of the 25th European Regional Conference of the International Telecommunications Society, Belgium, June 2014.

[11] "How artificial intelligence (AI) is helping to make the smart cities concept a reality," March 2019, https://achievion.com/blog/how-artificial-intelligence-ai-is-helping-to-make-the-smart-cities-concept-a-reality.html

[12] **F. Cugurullo**, "Urban artificial intelligence: From automation to autonomy in the smart city," Frontiers in Sustainable Cities, July 2020.

[13] **N. Graham** and M. **Sobiecki**, "Artificial intelligence in smart cities," May 2020 http://www.businessgoing.digital/artificial-intelligence-in-smart-cities/

[14] **X. Guo et al.**, "Review on the application of artificial intelligence in smart homes," Smart Cities, vol. 2, 2019, pp. 402–420.

[15] **P. Coppola** and **F. Cheli**, "How AI can help cities to better manage transport," October 2020, https://cities-today.com/industry/how-ai-can-help-cities-to-better-manage-transport/

[16] "AI opens new avenues for smart cities," https://sciencebusiness.net/data-rules/news/ai-opens-new-avenues-smart-cities

[17] **V. S. Bisen**, "How AI can be used in smart cities: Applications role & challenge," https://medium.com/vsinghbisen/how-ai-can-be-used-in-smart-cities-applications-role- challenge-8641fb52a1dd

[18] **C. G. Kirwan** and **F. Zhiyong**, Smart Cities and Artificial Intelligence: Convergent Systems for Planning, Design, and Operations. Elsevier, 2020.

[19] **A. Picon**, Smart Cities: A Spatialised Intelligence. John Wiley & Sons, 2015.

[20] **M. Hatti (ed.)**, Renewable Energy for Smart and Sustainable Cities. Springer, 2018.

[22] **K. Lyu**, AI-Based Services for Smart Cities and Urban Infrastructure. IGI Global. 2020.

CHAPTER 3
AI IN SMART GRID

"I also believe that we have an extraordinary opportunity for the United States and European Union to lead the world in developing and implementing new and more efficient technologies - smart electrical grids and electrical vehicles." - *Hillary Clinton*

3.1 INTRODUCTION

We all depend on energy to do things. But energy is a limited resource which faces additional challenges of efficiency, sustainability, and de-carbonization. Electricity has been one of the most important and the most widely used forms of energy since the 19th century. Without a doubt, electricity is the blood of industrialization, while network connections are the nerves. The electric power has changed our society. The electric industry essentially consists of four primary functional areas: generation, transmission, distribution, and utilization. This entire system is popularly known as the "grid." Thus, the power grid is a dynamic power system that delivers electric power from a generation system through transmission and distribution systems to end-users.

Traditionally, power grids were designed as one-directional systems sending power to factories, offices, and homes. The power infrastructure consists of a vast network of power plants, transmission lines, and distribution centers (comprising roughly 5,800 power plants and over 2.7 million miles of power lines). This deteriorating structure supports one-way power flow from centralized generation to end customers and is yet to receive a modern overhaul [1]. The energy sector worldwide faces some challenges related to rising demand of increasing global population, integration with various distributed components, efficiency, changing supply and demand patterns. These challenges are more acute in developing nations.

The smart grid (SG) has emerged as a replacement for ill-suited power systems in the21st century. As illustrated in Figure 3.1, the smart grid is an electrical power grid that is integrated with an AI-enabled, two-way communication network providing energy and information.

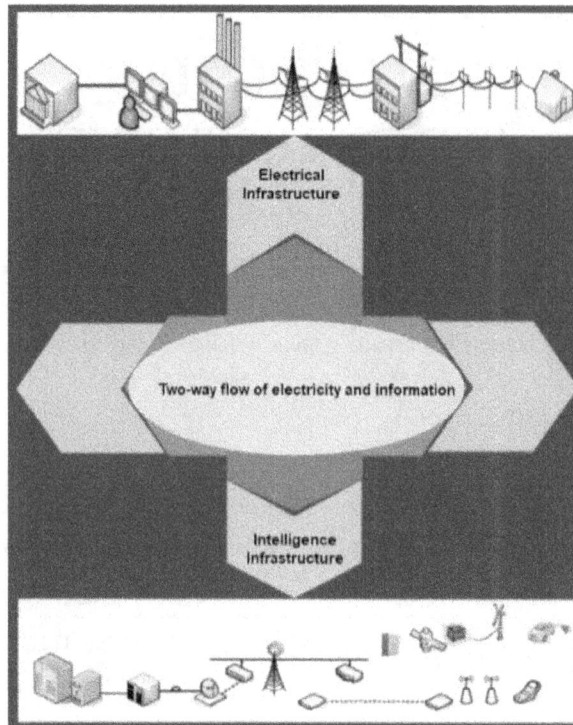

Figure 3.1 Data and electricity flows in Smart Grids.

The word "smart" in smart grid (also known as the next-generation power grid) refers to the notion of a power grid with intelligence. The main objective of the smart grid is to bring reliability, flexibility, efficiency, and robustness to the power system. Smart grid does this by introducing two-way data communications into the power grid. Thus, the smart grid consists of the power infrastructure and communication infrastructure, which correspond to the flow of power and information respectively [2]. It is a technology that enables instantaneous feedback from various sensors and devices on the power grid. The concept of smart grid is shown in Figure 3.2.

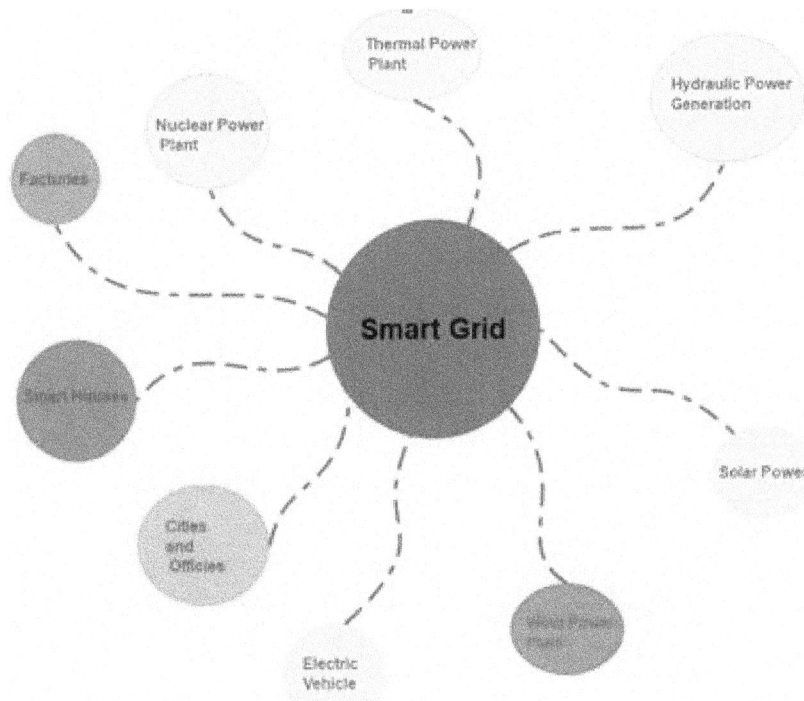

Figure 3.2 The concept of smart grid.

Smart grid technologies such as microgrids, generators and solar panels may help curb climate change. They can also balance peak demand and make energy generation sources more efficient [3].

This chapter investigates the role of AI in the different sectors of the smart grid. It begins by giving a brief review of artificial intelligence. It discusses AI in smart grid, its enabling technologies, actual motivation for AI, and its various applications. It highlights the benefits and challenges of incorporating AI in smart grid. The last section concludes with comments.

3.2 REVIEW OF ARTIFICIAL INTELLIGENCE

Artificial Intelligence is the intelligence of machines, algorithms, programs, or systems. It is one of the most efficient and potential technologies of all time. It is a subdiscipline of computer science that aims at enabling machines to perform tasks intelligently. AI has become one of the fastest growing technologies and is expected to play an important role in energy, transportation, healthcare, cybersecurity, education, etc.

The term "artificial intelligence" (AI) was first used at a Dartmouth College conference in 1956. AI is now one of the most important global issues of the 21st century. AI is the branch of computer science that deals with designing intelligent computer systems that mimic human intelligence, e.g., visual perception, speech recognition, decision-making, and language translation. The ability of machines to process natural language, to learn, to plan makes it possible for new tasks to be performed by intelligent systems. The main purpose of AI is to mimic the cognitive function of human beings and perform activities that would typically be performed by a human being. Without being taught by humans, machines use their own experience to solve a problem.

AI is a collection of programmed algorithms to mimic human decision-making. The components of AI include:

- Expert systems
- Fuzzy logic
- Neural networks
- Machine learning

- Deep learning
- Natural Language Processors
- Robots

All AI systems rely on algorithms, which are basically a set of instructions on how to organize and manage data. AI is well-suited to form the analytical foundation of smart city programs [4]. Although AI is in its early stages of implementation, it is poised to revolutionize the way we produce, transmit, and consume energy. AI will be the brain of future smart grid, which is the basis of smart energy.

3.3 AI IN SMART GRID

Although AI is relatively new, it is poised to revolutionize the way we produce, transmit, and consume energy. AI will constitute the brain of future smart grid. The power sectors have started to use AI and related technologies for communication between smart grids, smart meters, and Internet of things devices.

Like other industries, the power sector has been inundated by AI-enabled technologies. To support the existing systems and to extend the flexibility and applicability of smart grids, AI has been naturally adapted. As a result, large regional grids will be replaced by microgrids that manage local energy demand. Researchers at Argonne National Laboratory (the first US national laboratory) are developing new ways to extract insights from the massive data on the electric grid, with the intent of ensuring greater reliability, resilience, and efficiency. They are working on optimization models that use machine learning to simulate the electric system and the severity of various problems [5]. Some of the adapted AI techniques in the smart grid include [6]:

- Managing the grid users and controllers
- System based operation strategies for the grid
- Power supply optimization
- Consensus-based intelligent distribution techniques

- Machine learning and deep learning enabled costing mechanisms
- Intelligent energy storage systems
- Intelligent voltage profile regulation techniques using smart algorithms
- Integrating privacy into the smart grid.

From NIST, the eight priority areas for standardization of the smart grid are [7]:

1. Demand response and consumer energy efficiency: Targets numerous customer segments to involve them in making efficient energy consumption by controlling and scheduling their consumption pattern.

2. Wide-area situational awareness (WASA): Provides the network operators accurate information at the right time to make appropriate decisions.

3. Energy storage: Stores energy for later use to facilitate consumers with cheaper electricity. It provides more flexibility and helps to balance the grid by providing backup to the intermittent renewable energy sources.

4. Electric transportation: Provides economical energy, saves the environment, enhances living standards, and drives economic growth via various electric vehicles, e.g., plug-in electric vehicles (PEVs), battery electric vehicles (BEVs), plug-in hybrid electric vehicles (PHEVs).

5. Network communications: Integrates smart energy components via bidirectional communication channels.

6. Advanced metering infrastructure (AMI): Gathers and analyzes information from smart meters and provides efficient/intelligent management opportunities to the consumers.

7. Distribution grid management: Improves the stability of the grid and reduces the losses.

8. Cybersecurity: Protects data collected from the smart grid via ICT from various cyber-attacks.

3.4 ENABLING TECHNOLOGIES

The smart grid resembles the Internet in many aspects. The development of the smart grid is a natural consequence of the proliferation of AI, IoT, ICT, and 5G technologies.

- **Artificial Intelligence (AI):** AI is the backbone of a true smart grid. The development of advanced technologies using state-of-the-art AI to deal with problems in the smart grid is essential. AI will play the crucial role of taking into account the millions of variables and data points, including weather, demand, location, generation assets, etc. It will proactively decide for every home where the power will come from and how much it will cost.

- **Internet of Things (IoT):** Through the automation of various processes, the use of IoT introduces unprecedented levels of comfort and better control of the devices around us. Such is the case of the smart grid – a technology that enables instantaneous feedback from various sensors and devices on the operation of the power grid by streamlining the power delivery process. There has been general consensus that the IoT will control every vehicle, every home, every factory, every generator, every building, etc. in an integrated way.

The combination of AI and IoT is transforming the way smart grids are developed and used. The smart grid is poised for major change through these technologies. The technologies will also modernize the smart grid.

3.5 APPLICATIONS OF AI IN SMART GRID

Artificial Intelligence is everywhere. It is the fastest growing branch of the high-tech industry. AI's vast potential has motivated several initiatives by utilities, government agencies, and academia. The US Department of Energy (DOE) recently established a new Artificial Intelligence and Technology Office (AITO) to coordinate the agency's AI development, delivery, and adoption [4]. In September 2017, the DOE funded researchers at Stanford University to use artificial intelligence to improve grid stability. The UK's National Grid collaborated with DeepMind has added AI technology to the country's electricity system. The German government sees AI as a key strategy for mastering some of our greatest challenges such as climate change and pollution.

AI is becoming more and more important in the energy industry. Typical areas of application are autonomous grids, smart meters, energy consumption, electricity trading, failure management, and energy storage. Figure 3.3 shows some of the applications of AI in energy industry. These applications are discussed as follows [8-11].

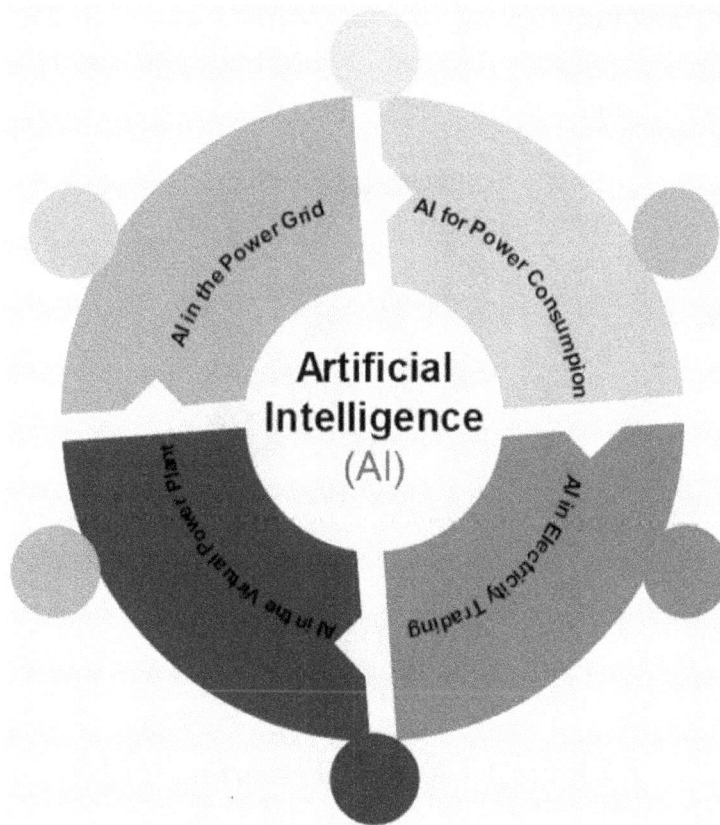

Figure 3.3 Some applications of AI in energy industry.

Autonomous Grid: In the US, the DOE is developing an autonomous grid using AI. With the power grids now collecting energy from different sources and the increasing decentralization of the grids, operating the grids has become more complex. This requires analyzing massive data. Artificial intelligence helps process this data as quickly and efficiently as possible, thereby bringing stability and efficiency.

Smart Meters: A smart meter is a high-tech meter that measures electricity consumption and provides additional information to the utility company unlike the conventional, analog meter. Smart meters are physically located within the building Smart meters (SMs) are essentially digital meters that read remotely over a secure wireless network. They are regarded smart in that consumers can moderate their energy consumption based on generated electricity information. They are an important component of the smart grid system. They will be able to constantly monitor demand and supply of customers. These smart meters process information that can be related to a person and be privacy sensitive. With smart metering, one can monitor every appliance and provide the homeowner with a comprehensive picture of their energy usage [12]. The European Union is strongly pushing the adoption of the smart grid and smart meters. Figure 3.4 shows a typical smart meter [13].

Figure 3.4 A typical smart meter [14].

Energy Consumption: In addition to making the power grid smart, flexible, and me so that greater optimize user's behavior and manage energy consumption. In a smart networked home, the networked devices react to prices on the electricity market and adapt to household usage accordingly. By monitoring the energy consumption pattern of individuals and businesses, AI companies can offer solutions to optimize usage, save electricity, and reduce costs. For example, the Smart thermostat Nest adapts to temperature
according to user behavior to reduce energy consumption.

Electricity Trading: In electricity trading, AI helps improve forecasts. Machine learning and neural networks play an important role in improving forecasts in the energy industry.

Failure Management: Without regular checks on power equipment, equipment failures are common. Using AI to observe equipment and detect failures before they happen can save money, time, and lives. Also, the use of AI helps not only in detecting a failure but also in isolating the faulty subsystem so that greater.

Energy Storage: The smart grid with energy storage will continuously collect massive data to make timely decisions on how best to allocate energy resources. Combined with other technologies such as big data, the cloud and the Internet of things (IoT), energy storage with AI can play an important role in power grid management. A smart grid with energy storage is able to use energy sources in the most efficient way by better integrating renewable resources.

Electric Vehicles: The use of electric vehicles (EVs) and high-speed electric trains will have to become widespread in order to reduce our reliance on oil for transportation. The coming years are likely to see the large-scale EV adoption that will shift the energy requirements of transport from fossil fuels to renewable electricity from the smart grid. EVs place a considerable additional load on the grid due to the high charging rates. The use of EVs is a key component in achieving overall sustainable solutions for utilities, cities, and nations. A typical electric car is shown in Figure 3.5.

Figure 3.5 A typical electric car.

These applications have achieved impressive results. They are simply a foretaste of what is ultimately possible. There are many more applications of AI in smart grid or energy industry such as energy management, network planning, fraud detection, load forecasting, stability analysis, security assessment, stability assessment, fault diagnosis, fault prediction, and stability control in smart grids.

3.5 BENEFITS

Smart grid is currently the new trend for clean, sustainable, efficient, and reliable energy generation and delivery. Its aim is to provide grid stability, balance supply and demand and ensure reliability of power supply. A key requirement for a safe and efficient smart grid is that supply and demand are always in perfect balance. The emerging artificial intelligence tools enable stability analysis and control in smart grids.

The application of AI-based technologies to the power grid cuts energy waste, facilitates the use of clean and renewable energy sources, and improve the planning, operation, and control of the power systems. The technologies can also help improve power management, efficiency, stability, resilience. and transparency, and increase the use of renewable energy sources. Smart grid facilitates large amounts of renewable energy integration. It is the foundation that will enable the adoption of EVs in the marketplace [14].

A major benefit of AI is the ability for the customers and the grid to be connected directly, creating win-win situation. The price of solar panels has come down recently years to bolster the cost-effectiveness of renewables. This can lead to a more efficient market and more cost-effective electricity production. It is realistic to expect the smart grid system to lower electricity bills and prevent catastrophic blackouts. AI can help use less energy to accomplish more. It is also helping compress and analyze the massive amounts of data produced by energy industry.

3.6 CHALLENGES

In most nations, the electric power grid has changed very little since it was first installed. The grid itself relies on ageing infrastructure and has significant inefficiencies arising from losses within the transmission and distribution networks. The notion of a power grid that makes extensive use of renewable generation challenges this predicament. It presents many challenges in terms of power systems engineering, telecommunications, and cybersecurity.

Smart grid faces a wide range of other challenges, such as extreme weather, imperfections in the available infrastructure, reliability issues, equipment failures, gigantic customer base, decentralized generation, and decarbonizing the global economy. A major challenge is the rise of distributed generation, where individual customers generate and use their own electricity from renewable sources, such as wind and solar. The current power system was not designed to accommodate this diversification in energy sources and fluctuating supplies of renewable energy. Industry leaders in the AI energy grid industry are aware of these challenges and must address them. The use of AI in smart grid is not without risks. One risk has to do with the privacy of customer data collected by AI systems. For example, smart meter information can reveal presence, sleeping habits, use of devices, etc. This information is not limited to staying home because smart meters can be read at a distance.

3.7 THE FUTURE OF AI IN SMART GRID

Artificial intelligence holds significant potential across a wide array of sectors. It is expanding its scope over tasks traditionally performed by humans. Its applications in the energy industry are helping to make the grid "smarter" or more responsive, thereby develop the smart grid of the future. AI has been regarded as the brain behind the future smart grid, enabling the real-time optimization and automation of distribution planning and operation decision. The smart grid is a dynamic system that continues to evolve as technologies are tested and perfected [16].

Although the use of AI in the smart grid faces some challenges such as insufficient reliability, imperfect infrastructure, and lack of special algorithm for power industry, AI is a powerful tool to push smart grid into the new generation of power systems. AI supports and optimizes electric networks around the world, pushing the concept closer towards a global adoption. It should be added to all levels in the energy grids, in order to enhance their development.

Although the "smart" is not quite here yet, it is slowly becoming a reality. It is on its way to usher the energy industry into a new era of reliability, availability, and efficiency. Significant investments in the infrastructure will be needed to help smart grids fully take off [17].

New generation of energy networks will make efficient use of renewable energy sources, support real time and efficient demand response, as well as the large-scale deployment of electric vehicles (EVs) [18]. To provide a low-cost and flexible solution for the grid-wide information exchange, wireless communications technology is expected to play the key role in the emerging smart grid applications.

3.8 CONCLUSION

The smart grid is the developmental trend of power systems. It is evolving to become more flexible, sustainable, and distributed. It has attracted much attention all over the world. It provides a platform for clean, sustainable, efficient and reliable energy generation, delivery, and consumption. It is inevitable that the smart grid will become part of our society.

The smart grid is slowly becoming a reality. The technology is on its way to revolutionize power supply. There is need for consumer education, development of standards and regulations, and information sharing between projects [19]. The global energy demand is expected to increase steadily in the future. Therefore, future generations should realize that AI and energy are not mutually exclusive career paths. For more information about artificial intelligence in smart grid, one should consult the books in [20,21] and a related journal: IEEE Transactions on Smart Grid.

REFERENCES

[1] **F. Wolfe**, "How artificial intelligence will revolutionize the energy industry," http://sitn.hms.harvard.edu/flash/2017/artificial-intelligence-will-revolutionize-energy- industry/

[2] **M.N.O. Sadiku**, and **S.M. Musa** and **S. R. Nelatury**, "Smart grid – An introduction," International Journal of Electrical Engineering & Technology, vol. 7, no.1, Jan-Feb, 2016, pp. 45-49.

[3] **V. Robu**, "Why artificial intelligence could be key to future-proofing the grid," https://robohub.org/why-artificial-intelligence-could-be-key-to-future-proofing-the-grid/

[4] **A. Tomer**, "Artificial intelligence in America's digital city," July 2019, https://www.brookings.edu/research/artificial-intelligence-in-americas-digital-city/

[5] **C. Nunez**, "Artificial intelligence can make the U.S. electric grid smarter," June 2019, https://www.tdworld.com/grid-innovations/smart-grid/article/20972769/artificial intelligence-can-make-the-us-electric-grid-smarter

[6] **V. Sooriarachchi**, "Prospect of adapting artificial intelligence in smart grids for developing countries," https://smartgrid.ieee.org/newsletters/november-2020/prospect-of-adapting-artificial-intelligence-in-smart-grids-for-developing-countries

[7] **S. S. Ali** and **B. J. Choi** "State-of-the-art artificial intelligence techniques for distributed smart grids: a review," Electronics, vol. 9, 2020.

[8] "What is artificial intelligence in the energy industry?" https://www.next-kraftwerke.com/knowledge/artificial-intelligence

[9] "5 Ways the energy industry is using artificial intelligence," March 2018, https://www.cbinsights.com/research/artificial-intelligence-energy-industry/

[10] **S. Bilodeau**, "Artificial intelligence in a 'no choice but to get it smart' energy industry!" April 2019, https://towardsdatascience.com/artificial-intelligence-in-a-no-choice-but-to-get-it-smart-energy-industry-1bd1396a87f8

[11] **S. D. Ramchurn et al.**, "Putting the 'smarts' into the smart grid: a grand challenge for artificial intelligence," Communications of the ACM , vol. 55, no. 4, April 2012.

[12] **M. N. O. Sadiku, S.M. Musa, A. Omotoso**, and **A.E. Shadare**, "A primer on smart meters," International Journal of Trend in Research and Development, vol. 5, no. 4, 2018, pp. 65-67.

[13] "Smart meter," Wikipedia, the free encyclopedia https://en.wikipedia.org/wiki/Smart_meter

[14] **E. F. Merchant**, "The next big obstacle for electric vehicles? Charging infrastructure," November 2017, https://www.greentechmedia.com/articles/read/the-next-big-obstacle-for-electric-vehicles-charging-infrastructure

[15] **C. Frye**, "4 Ways artificial intelligence is powering the energy industry," November 2018, https://www.kolabtree.com/blog/4-ways-artificial-intelligence-is-powering-the-energy- industry/

[16] **M. Sulikowski**, "Smart grid - AI at the service of the power distribution network," June 2019, https://naturaily.com/blog/smart-grid-ai-in-power-distribution-network#:~:text=The%20term%20%E2%80%9Csmart%20grid%E2%80%9D%20describes,faster%20restoration%20after%20power%20blackouts.

[17] **N. Bassiliades** and **G. Chalkiadakis**, "Artificial intelligence techniques for the smartgrid," Advances in Building Energy Research, vol. 12, no. 1, 2018, pp. 1-2.

[18] **L. A. Kumar**, **L. S. Jayashree** , and **R. Manimegalai (eds**.), Proceedings of International Conference on Artificial Intelligence, Smart Grid and Smart City Applications. Springer, 2020.

[19**] M. Sulikowski**, "Smart grid - AI at the service of the power distribution network," June 2019, https://naturaily.com/blog/smart-grid-ai-in-power-distribution-network#:~:text=The%20term%20%E2%80%9Csmart%20grid%E2%80%9D%20de scribes,faster%20restoration%20after%20power%20blackouts.

[20] **L. A. Kumar, L. S. Jayashree, and R. Manimegalai** (eds.), *Proceedings of International Conference on Artificial Intelligence, Smart Grid and Smart City Applications.* Springer, 2020.

[21] **S. F. Bush**, Smart Grid: Communication-Enabled Intelligence for the Electric Power Grid. John Wiley & Sons, 2014.

CHAPTER 4
AI IN HEALTHCARE

"AI is only as good as the humans programming it and the system in which it operates. If we are not careful, AI could not make healthcare better, but instead unintentionally exacerbate many of the worst aspects of our current healthcare system." — *Bob Kocher*

4.1 INTRODUCTION

Recently, we have witnessed a wave of emerging technologies, from Internet of things and blockchain to artificial intelligence (AI), demonstrate significant potential to transform and disrupt multiple sectors, including healthcare. Healthcare is shifting from traditional hospital-centric care to a more virtual care that leverages the latest technologies around artificial intelligence, deep learning, big data, genomics, robotics, increased access to data, additive manufacturing, and wearable and implanted devices [1].

Today, artificial intelligence (AI) is shorthand for any task a machine can perform just as well as, if not better than, humans. AI represents the hopes and fears of an industry seeking more intelligent solutions. AI is an interdisciplinary field covering numerous areas such as computer science, psychology, linguistics, philosophy, and neurosciences. The central objectives of AI research include reasoning, knowledge, planning, learning, natural language processing, perception, and the ability to move and manipulate objects [2].

Although AI is a branch of computer science, there is hardly any field which is unaffected by this technology. Common areas of applications include agriculture, business, law enforcement, oil and gas, banking and finance, education, transportation, healthcare, engineering, automobiles, entertainment, manufacturing, speech and text recognition, facial analysis, and telecommunications [3]. Healthcare is regarded as the next domain to be revolutionized by AI. In healthcare, AI can help manage and analyze data. AI can have a significant impact in making healthcare more accessible, especially in developing countries, where shortages of healthcare practitioners are most severe. There are many cases in which AI can perform healthcare tasks as well or better than humans.

This chapter provides an overview of a broad range of applications of AI in healthcare. It begins by reviewing AI and its enabling technologies. It covers a variety of applications of AI in healthcare. It discusses some international trends on the use of AI in healthcare. It highlights the benefits and challenges of AI in healthcare. The last section concludes with comments.

4.2 REVIEW ON ARTIFICIAL INTELLIGENCE

Artificial intelligence is the ability of a computer to function appropriately and with foresight in its environment. The term "artificial intelligence" (AI) was coined in 1956 by John McCarthy during a conference held on this subject. AI is the branch of computer science that deals with designing intelligent computer systems that mimic human intelligence. The ability of machines to process natural language, to learn, and to plan makes it possible for new tasks to be performed by intelligent systems. The main purpose of AI is to mimic the cognitive function of human beings and perform activities that would typically be performed by a human being. AI is stand-alone independent electronic entity that functions much like human healthcare expert. Today, AI is integrated into our daily lives in several forms, such as personal assistants, automated mass transportation, aviation, computer gaming, facial recognition at passport control, voice recognition on virtual assistants, driverless cars, companion robots, etc. [4]. AI technologies are performing better and better at analyzing health data, thereby helping doctors better understand the future needs of their patients.

An important feature of AI technology is that is can be added to existing technologies. AI has benefited many areas such as chemistry and medicine, where routine diagnoses can be initiated by AI-aided computers. It embraces a wide range of disciplines such as computer science, engineering, machine learning, chemistry, biology, physics, astronomy, neuroscience, and social sciences.

AI is not a single technology but a range of computational models and algorithms. The major disciplines in AI include expert systems, fuzzy logic, and artificial neural networks (ANNs), machine learning, deep learning, natural language processing, computer vision, and robotics. These tools or technologies have been used to achieve AI's goals.

4.3 ENABLING TECHNOLOGIES

The technologies enabling the use of AI in healthcare include AI, machine learning, deep learning, robotics, and big data. They are the technologies transforming healthcare. These enabling technologies have immediate relevance to the healthcare field, but the specific processes and tasks they support vary [5-7].

• **Artificial intelligence (AI):** This is the use of computer science to develop a machine that can be trained to learn, reason, communicate, and make human-like decisions. It is a technology that is rapidly being adopted in many industries to improve performance, precision, time efficiency, and cost reduction. The use of artificial intelligence in healthcare is an emerging scientific area that aims to generate healthcare intelligence by analyzing health data. AI is becoming increasingly attractive in healthcare industry and changing the landscape of healthcare and biomedical research.

• **Machine learning:** This is a statistical technique for fitting models to data and to "learn" by training models with data. Machine learning (ML) is one of the most common forms of AI. The great majority of machine learning applications in healthcare requires a training dataset. This is called supervised learning. Unsupervised learning involves the ability to identify previously undiscovered predictors and learning associations in data. The most complex forms of machine learning involve deep learning (DL) The relationship between AI, ML, and DL is shown in Figure 4.1.

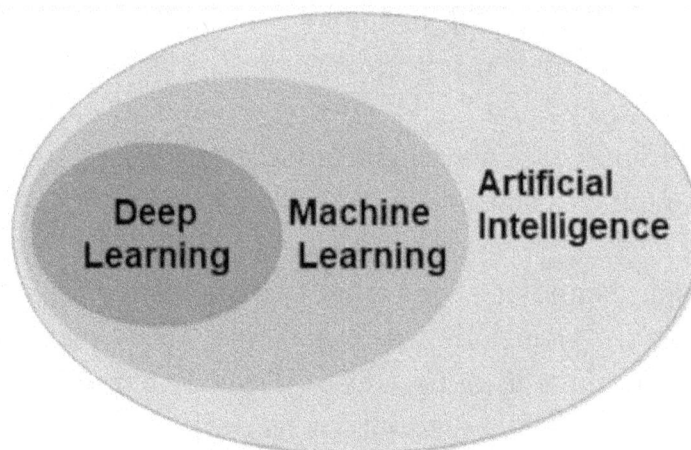

Figure 4.1 The relationship between AI, ML, and DL.

- **Robotics:** Given the successful performance of robotics in a wide range of industries, from vehicle manufacturing to space exploration, robots have been introduced to transform a healthcare procedure like a surgery into an assisted operation definitely safer and more convenient for both doctors and patients. Robots have been used in healthcare for more than 30 years. Over the years, a wide range of robots is developed to serve different purposes within the healthcare environment. This results in various kinds of healthcare robots such as surgical robots, logistics robots, disinfectant robots, cleaning robots, pill robots, laboratory robots, rehabilitation/exoskeleton robots, nursing robots, telepresence robots, therapy robots, assistive robots, robotic prosthetic limbs, diagnostics robots, and many other types. Robots have the potential to revolutionize end of life care with its increasing ability to translate the complexity in data into actionable clinical decisions.

4.4 APPLICATIONS OF AI IN HEALTHCARE

AI has numerous applications in healthcare. Some of these applications are shown in Figure 4.2.

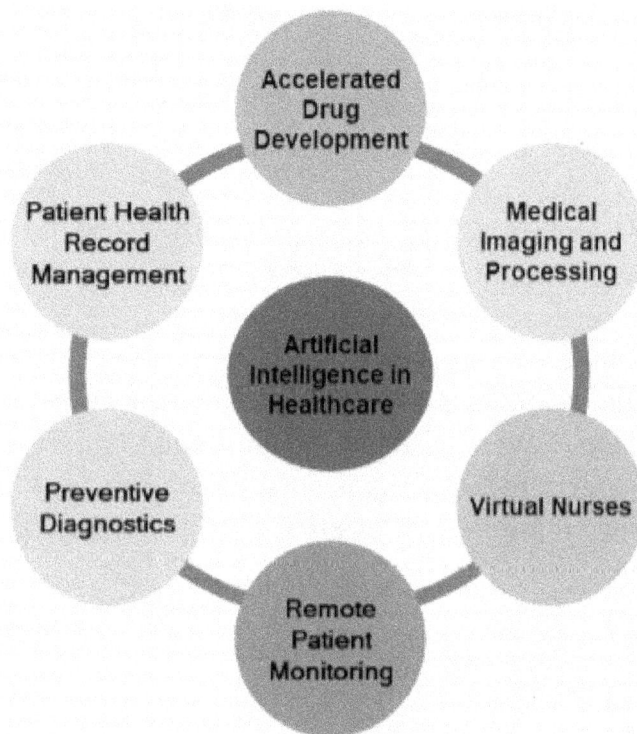

Figure 4.2 AI in healthcare [8].

The primary aim of AI applications in the health domain is to analyze relationships between prevention or treatment techniques and patient outcomes. Various AI applications have been developed to solve some of the most pressing problems that currently face healthcare industry. The research focus is mainly around a few disease types: cancer, nervous system disease, and cardiovascular disease. The following specialties in medicine have shown an increase in research regarding AI [8-12].

- **Radiology:** The ability to interpret imaging results with radiology may aid clinicians in detecting a minute change in an image that a clinician might not notice otherwise. This is the widest application of AI in medicine, but providers are just beginning to tap into the potential of what AI technology has to offer. The practice of radiology relies primarily on imaging for diagnosis and is very amenable to deep-learning techniques. As AI continues to expand in its ability to interpret radiology, it may be able to diagnose more people with the need for less doctors as there is a shortage in many nations. The emergence of AI technology in radiology is perceived as a threat by some specialists. Today, AI algorithms are outperforming radiologists at spotting malignant tumors.

- **Oncology/Cancer:** In breast cancer diagnosis and the detection of lung cancer, AI algorithms have been shown to be better and more effective than a human. An AI-assisted robots can analyze data from pre-op medical records and guide the surgeon's instrument in real-time during a procedure. It has been demonstrated that the IBM Watson for oncology would be a reliable AI system for assisting the diagnosis of cancer. (IBM has also developed Watson for Health and Watson for Genomics.) The popularity of robot-assisted surgery in hospitals is skyrocketing since it has led to fewer surgery-related complications. Figure 4.3 shows that robotic surgery offers more precision [13].

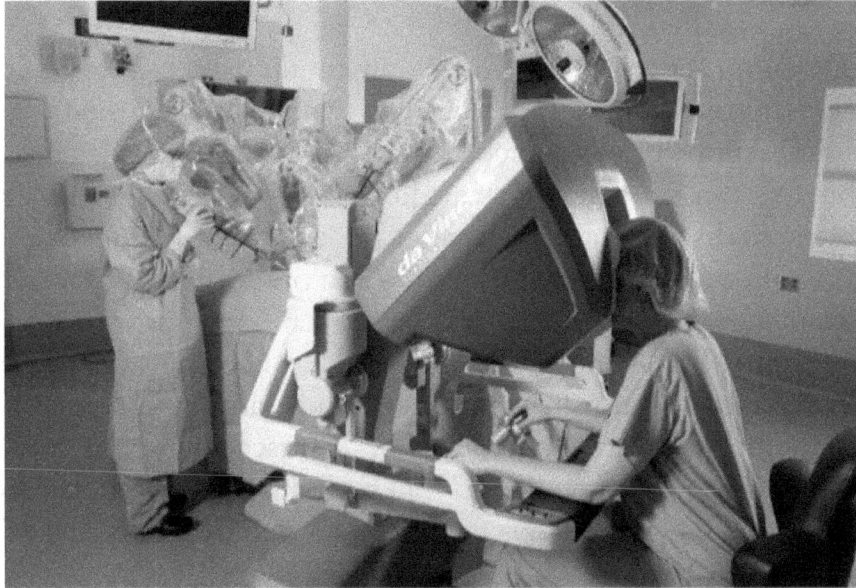

Figure 4.3 Robotic surgery offers more precision [13].

A prime example of minimally invasive surgery is the one performed by da Vinci Si Robot. Numerous hospitals are equipped with such robotic surgery facilities and surgeons' job is reduced to playing the video game from a distance.

- **Dermatology:** This is an imaging abundant specialty. There is a natural fit between the dermatology and deep learning. Most applications of AI in dermatology are aimed at diagnosing and preventing the onset of skin disease. AI has the potential to play a major role in supporting the decision-making of dermatologists on skin conditions and specific cancers. Radiology, pathology, and dermatology are anticipated to be the first clinical specialties to experience large-scale change due to the incorporation of AI into work practices.

- **Telemedicine:** Telemedicine (also known as telehealth or chealth) may be regarded as the transmission of medical images between healthcare centers for diagnosis across distance. It allows healthcare practitioners to diagnose, treat, and monitor patients at a distance using telecommunications technology. Telemedicine is used in a variety of specialties including radiology, neurology, and pathology. The ability to monitor patients using AI may allow for the communication of information to physicians, if possible, disease activity may have occurred [14,15].

- **Drug Development:** The drug development industry is bogged down by skyrocketing development costs and research. Due to breakthroughs in technology, biopharmaceutical companies are quickly taking notice of the efficiency, accuracy and knowledge that AI can provide. AI may help assist researchers to identify genetic mutations that cause disease and help predict the effects of treatments. Application of AI has drawn significant interest in the analysis and processing of biological and genetic information resulting in accelerated drug discovery and identification of selective molecular modulators. Prediction of this behavior has become an easier task, thanks to AI. Today development of novel hypotheses and treatment strategies, prediction of progression, and the associated outcomes of diseases and even evaluating of the pharmalogical profiles of drug conditions – all such crucial steps in healthcare are being routinely addressed by AI. These steps facilitate a procedure called target fishes. It helps immensely in drug delivery systems.

- **Electronic Health Records:** Electronic health records are essentially the digital version of a patient's medical history. They are crucial to the digitalization and information spread of the healthcare industry. They contain the clinical history of patients and could be used to identify the individual risk of developing cardiovascular diseases, diabetes, and other chronic conditions. Using an AI tool to scan EHR data can accurately predict the course of disease in a patient.

- **Mobile Health:** Mobile health (or mHealth) refers to the practice of medicine via mobile devices such as mobile phones, tablet computers, personal digital assistants (PDAs), and wearable devices. It has emerged as the creative use of emerging mobile devices to deliver and improve healthcare practices. It integrates mobile technology with the health delivery with the premise of promoting a better health and improving efficiency. mHealth benefits immensely from AI. AI algorithms, sensor technology, and advanced data are helping transform smartphones into full health-management platforms. The evolution of mHealth can be seen in the improved availability of healthcare services, increased efficiency in the treatment process, reduced costs, and the creation unprecedented opportunities for preventive care. mHealth assistants will become a popular alternative in developed countries, where doctors are very busy [16,17].

- **Medical Research:** AI can be used to analyze and identify patterns in large and complex datasets. It can also be used to search the scientific literature for relevant articles. AI systems used in healthcare could also be valuable for medical research by helping to match suitable patients to clinical studies. AI can aid early detection of infectious disease outbreaks and sources of epidemics. AI has also been used to predict adverse drug reactions [18].

- **Consumer Wearables:** Healthcare is a unique industry that can get a huge amount of benefits from both AI and wearable technology. The integration of wearable technology in healthcare with technologies such as AI, augmented reality (AR), and virtual reality (VR) makes the wearable devices smarter. Diagnostic wearables technology can be used to diagnose various diseases in real-time. The combination of consumer wearables with AI is being applied to enable caregivers to better monitor and detect potentially life-threatening episodes at earlier stages.

AI can also be used in neurology, cardiology, stroke, aging, health surveillance health, health monitoring, health insurance, hospital inpatient care, healthcare management, urban healthcare system, suicide risk prediction, emergence medicine, detection of disease, management of chronic conditions, delivery of health services. The scope of possible applications of AI in healthcare are almost limitless. Future uses for AI include Brain-computer Interfaces (BCI) which will help those with trouble moving or speaking. Medical institutions, such as The Mayo Clinic, Massachusetts General Hospital, Memorial Sloan Kettering Cancer Center, and National Health Service, have developed AI algorithms for their departments. Major technology companies such as https://en.wikipedia.org/wiki/IBM IBM, Intel, Microsoft, and Google have also developed AI algorithms for healthcare [11].

4.5 INTERNATIONAL TRENDS

Artificial Intelligence has arrived in healthcare. The AI technology now moves towards globalization and it becomes necessary to track both government initiatives as well as regulatory changes around the world. There is global policy developments and investments in AI. Globally, health systems face many challenges: rising burden of illness, greater demand for health services, higher societal expectations, and increasing health expenditures. In spite of this, AI has been a strategic priority for governments around the world. The following are typical examples of international trends [19].

- **United States:** Like other industries, healthcare is witnessing a shift to consumerization, pushing payers and providers to focus on value-based care and improve the health outcomes. The US market has recently witnessed a relatively greater adoption of AI technology primarily to decrease the cost of care and improve the outcomes. Currently, the United States government is investing billions of dollars to promote the development of AI in healthcare. Companies are developing technologies that help healthcare managers improve business operations through increasing utilization. The Department of Health and Human Services (HHS) serves to enhance and promote the health and well-being of all Americans. HHS recognizes that AI will be a critical enabler of its mission and vision.

- **Canada:** Canada has a unique and time-limited opportunity to be world leaders in system design using AI technology. It has established itself as a world leader in AI technology-related research. Its main cities, Montreal, Toronto, and Vancouver, have become hubs for AI research and development, attracting companies like Google, Facebook, Uber, Microsoft, and Samsung. Canada aims at establishing responsible development of human-centric AI and facilitates international scientific collaboration [20].

- **United Kingdom:** The British government has announced its ambition to make the UK a world leader in AI and data technologies. The UK government has launched the Centre for Data Ethics and Innovation, as part of the UK's initiative to lead global governance on AI ethics.

- **France:** The government plans to establish France as leader in AI research. Its key initiatives include: (1) developing an open data policy to drive the adoption and application of AI in sectors like healthcare, (2) establishing a regulatory and financial framework to support the development of domestic "AI champions," and (3) putting in place regulation to ensure that AI developments remain transparent, explainable, and non-discriminatory.

- **China:** The government plans to develop intelligent and networked products such as vehicles, service robots, and identification systems. It announces investment in industry training resources, standard testing, and cybersecurity.

- **India:** India's AI strategy aims at promoting AI inclusion, an approach called "AIforAll." India is also attempting to establish itself as an "AI Garage", which allows the AI technology developed in India to be useable to the rest of world. India is rich in data due to the volume of patients. Artificial intelligence (AI) and machine learning (ML) are witnessing increasing adoption in the Indian healthcare setting. However, India is fraught with several problems like aging population, lack of adequate infrastructure, limited access to healthcare facilities, adherence to treatment, and availability of care providers [21].

AI solutions could improve access, quality, and efficacy of global health systems. A common trend is the international focus on the development of transparent and responsible AI policy.

4.6 BENEFITS

AI has unimaginable potential, and its benefits are enormous when implemented strategically. AI is increasingly developing the capability to do what humans do and do so more efficiently, more quickly, and at a lower cost. It will decrease medical costs as there will be more accuracy in diagnosis and better predictions in the treatment plan. Other benefits include [21,22]:

Improvements: The use of AI will deliver major improvements in quality and safety of patient care. It improves healthcare by preventing hospitalizations, reducing complications, decreasing administrative burdens, and improving patient engagement. AI tools have the potential to reduce administrative burdens and improve treatment.

Eliminates Repetitive Tasks: AI platform is designed to automate the healthcare Industry's most repetitive tasks, freeing up administrators to work on higher-level ones. AI is destined to drastically change clinicians' roles and everyday practices. AI has already begun making progress in healthcare by simplifying tedious and expensive procedures, guarding against human error, and promising to usher in a new era of patient care.

Decision-making: AI has been introduced as a technology to improve decision-making. Since AI makes decisions based on the data it receives as input, it is important that this data represents accurate information about patient.

Prediction: AI has been described in many ways. One way is to describe AI as "prediction technology." For example, predicting illness episodes that might be experienced in the future is an obvious application of AI. ML applications fundamentally perform some form of prediction. These applications demonstrate that AI can be a powerful tool for early and accurate diagnostics and even prediction of diseases.

Transform Healthcare: AI and ML technologies have the potential to transform and revolutionize healthcare by deriving new and important insights from the vast amount of data generated daily while delivering healthcare. The goal of healthcare is to become more personal, predictive, preventative, and participatory, and AI can make major contributions in these directions.

Increase Productivity: While a healthcare practitioner can treat one patient at a time, automated AI-powered health assistants can serve millions of patients simultaneously, thereby multiplying global productivity.

Minimizing Errors and Fraud: Errors and fraud are expensive in healthcare Organizations and also for insurers. Fraud detection is a time-consuming process that hinges on being able to quickly spot anomalies after the incident occurs in order to I ntervene. Health insurers are now using AI tools to search Medicare claims for patterns associated with reimbursement fraud.

Cost savings: Cost savings have been associated with AI applications involving robot-assisted surgery and virtual nursing assistants. Artificial intelligence should be used to reduce, not increase, health inequality. Some of these benefits of AI in healthcare is shown in Figure 4.4.

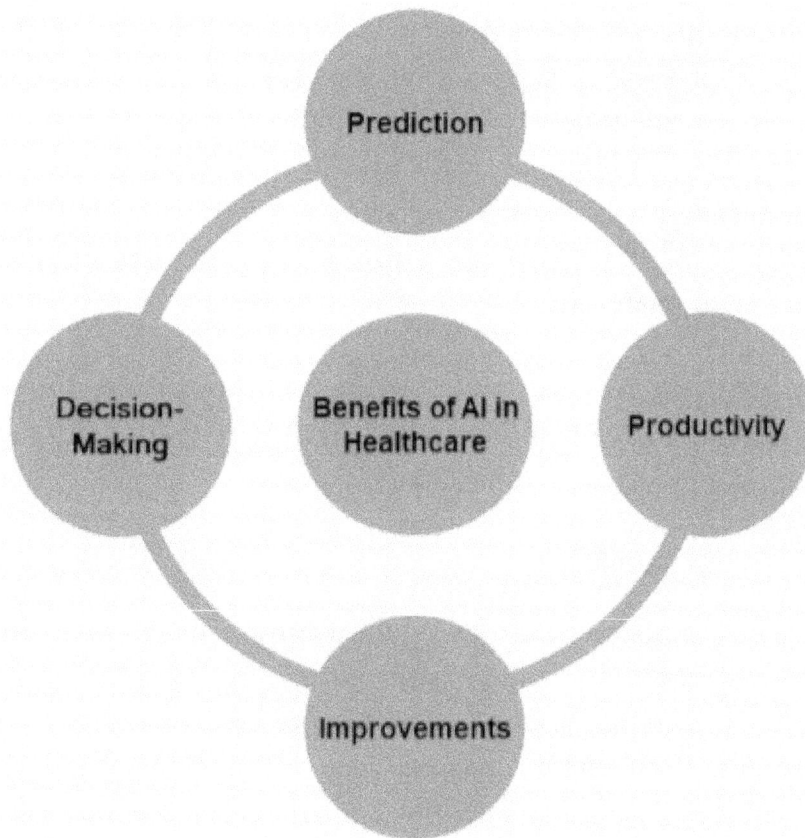

Figure 4.4 Benefits of AI in healthcare.

4.7 CHALLENGES

There are challenges hindering the successful AI technology adoption. While AI has achieved widespread adoption in certain sectors, the complexities of healthcare have resulted in slower adoption. Major challenges include [18,23,24].

- **Privacy and Security:** Healthcare providers must recognize that patient privacy and security must remain paramount. Therefore, AI companies should utilize valuable medical data while remaining compliant with law governing the protection of patient information and data ownership.

- **Fear of Job Loss:** There is concern that AI will lead to automation of jobs and substantial displacement of the workforce. This leads to the misconception that AI will replace human clinicians. As technology is increasingly implemented in workplaces, some fear that their jobs will be replaced by machines. It seems increasingly clear that AI systems will not replace human clinicians on a large scale. Doctors and nurses still have a number of unique and important advantages over AI. They can do a lot of things (touching, sensing, take a blood test, compassion, anxiety, memory, communication, learning, etc.) that an AI assistant cannot do because they are human traits that are difficult to model mathematically.

- **Regulation:** There is lack of regulations specifically for the use of AI in healthcare. Developing an adaptive AI technology in healthcare to meet regulatory requirements is an uphill task.

- **Ethics:** The social and ethical use of AI in healthcare presents significant challenges as some question about the ethical appropriateness of the use of AI. Ethical concerns in this area include professionalism, transparency, justice , safety, and data privacy, automation of jobs, and representation biases. We may likely encounter many ethical, medical, occupational and technological changes with AI in healthcare. Healthcare AI must be deployed in ways that promote quality of care and minimize potentially disruptive effects.

- **Lack of Trust:** Trust is a crucial factor influencing interactions between machines and humans. A lack of trust in the AI systems is a major hindrance in the adoption of this technology in healthcare. Trust in AI can be influenced by several factors such as user education, past experiences, user biases, and perception towards automation. Developing a healthy trust relationship in human-AI collaboration is a significant challenge. Promoting fairness through awareness can also improve trust.

Failure of hospitals to use AI prevents both the use of potentially life-saving technology and potential cost savings. Other significant challenges and issues include data preprocessing, consolidation, ubiquitous information, knowledge extraction, interpretability, and the need to ensure that the way AI is developed and used in a transparent and accountable manner, compatible with public interest.

4.8 CONCLUSION

The use of artificial intelligence in healthcare is evolving at a rapid rate. AI is penetrating every aspect of global healthcare. It has the potential to disrupt the healthcare industry. AI presents unprecedented opportunities in healthcare and major challenges for the patients, developers, providers, and regulators. Through our collective effort, AI can achieve all its lofty expectations to improve healthcare for patients across the world.

However, AI-based technologies are still quite controversial because they are not yet commonly used. In the near future, healthcare will be delivered as a seamless continuum of care and with a greater focus on prevention and early intervention. For more information about AI in healthcare, one should consult books in [26-34] and the related journals:

- Artificial Intelligence in Medicine
- Journal of Medical Artificial Intelligence.
- Journal of AI and Healthcare
- Future Healthcare Journal
- Journal of Healthcare Engineering
- Journal of Artificial Intelligence for Medical Sciences

REFERENCES

[1] **M. Wehde**, "Healthcare 4.0," IEEE Engineering Management Review, vol. 47, no. 3, Third Quarter, September 2019, pp. 24-28.

[2] **I. Sniecinskia** and **J. Seghatchianb**, "Artificial intelligence: A joint narrative on potential use in pediatric stem and immune cell therapies and regenerative medicine," Transfusion and Apheresis Science, vol. 57, 2018, pp. 422-424.

[3] **M. N. O. Sadiku**, "Artificial intelligence", IEEE Potentials, May 1989, pp. 35- 39.

[4] **Y. Mintz** and **R. Brodie**, "Introduction to artificial intelligence in medicine," Minimally Invasive Therapy & Allied Technologies, vol. 28, no. 2, 2019, pp. 73-81.

[5] **T. Davenport** and **R. Kalakota**, "The potential for artificial intelligence in healthcare," Future Healthcare Journal, vol. 6, no. 2, June 2019, pp. 94–98.

[6] "Artificial Intelligence, machine learning, and deep learning: Same context, different concepts." April 2018, https://master-iesc-angers.com/artificial-intelligence-machine-learning-and-deep-learning-same-context-different-concepts/

[7] **M. N. O. Sadiku, Y. Wang, S. Cui,** and **S.M. Musa**," Healthcare robotics: A primer," International Journal of Advanced Research in Computer Science and Software Engineering, vol. 8, no. 2, Feb. 2018, pp. 26-29.

[8] **D. Naik**, "AI in the healthcare world," August 2017, https://medium.com/@humansforai/ai-in-the-healthcare-world-88d13a815f35

[9] **F. Jiang et al.**, "Artificial intelligence in healthcare: Past, present and future," Stroke and Vascular Neurology, 2017.

[10] **R. O. Mason**, "Ethical issues in artificial intelligence," Encyclopedia of Information Systems, vol 2, 2003, pp. 239-258.

[11] "Artificial intelligence in healthcare," Wikipedia, the free encyclopedia
https://en.wikipedia.org/wiki/Artificial_intelligence_in_healthcare

[12] **M. N. O. Sadiku, T. J. Ashaolu**, and **S. M. Musa**," Artificial intelligence in medicine: A primer," International Journal of Trend in Research and Development, vol. 6, no. 1, Jan.-Feb. 2019, pp. 270-272.

[13] "Robotic surgery offers more precision, faster recoveries,"
https://news.sanfordhealth.org/general-surgery/robotic-surgery-precision-recovery/

[14] **M. N. O. Sadiku, M. Tembely**, and **S.M. Musa**," Telemedicine: A primer (Part 1),"
International Journal of Advanced Research in Computer Science and Software Engineering, vol. 9, no. 6, June 2019, pp.43-46.

[15] **M. N. O. Sadiku, M. Tembely**, and **S.M. Musa**," Telemedicine: Teleeverything phenomena (Part 2)," International Journal of Advanced Research in Computer Science and Software Engineering, vol. 9, no. 6, June 2019, pp.35-38.

[16] **M. N. O. Sadiku, A. E. Shadare**, and **S.M. Musa**," Mobile health," International Journal of Engineering Research, vol. 6, no. 11, Oct. 2017, pp. 450-452.

[17] **B. Dickson**, "How artificial intelligence is revolutionizing the mhealth industry,"
https://www.magzter.com/articles/1642/241037/59c9590a889f7

[18] **Nuffield Council on Bioethics**, "Artificial intelligence (AI) in healthcare and research"
http://nuffieldbioethics.org/wp-content/uploads/Artificial-Intelligence-AI-in-healthcare-and-research.pdf

[19] **S. E. Davies**, "Artificial intelligence in global health," Ethics & International Affairs, vol. 33, no. 2, Summer 2019.

[20] **A. Kassam** and **N. Kassam**, "Artificial intelligence in healthcare: A Canadian context," Healthcare Management Forum, 2019, pp. 1-5.

[21] **R. Mabiyan**, "How artificial intelligence can help transform Indian healthcare," ETHealthWorld, May 2018,
https://health.economictimes.indiatimes.com/news/health-it/how-artificial-intelligence-can-help-transform-indian-healthcare/64285489

[22] **G. Rong**, "Artificial intelligence in healthcare: Review and prediction case studies," Engineering, vol. 6, no. 3, March 2020, pp. 291-301.

[23] **O. Asan**, A. **E. Bayrak**, and **Avishek Choudhury**, "Artificial intelligence and human
trust in healthcare: Focus on clinicians," Journal of Medical Internet Research, vol 22, no 6, June 2020.

[24] **E. Crigger** and **C. Khoury**, "Making policy on augmented intelligence in health care," AMA Journal of Ethics, vol. 21, no. 2, 2019, pp. E188-191.

[25] "AI in healthcare: Keys to a smarter future,"
https://www.gehealthcare.com/-/media/b3a5e32538454cf4a61a4c58bd775415.pdf

[26] **P. Vasant (ed.),** Handbook of Research on Artificial Intelligence Techniques and Algorithms. Information Science Reference, 2015.

[27] **D. D. Luxton**, Artificial Intelligence in Behavioral and Mental Health Care. San Diego, CA: Elsevier, 2016.

[28] **A.** Panesar, Machine Learning and AI for Healthcare: Big Data for Improved Health Outcomes.Apress,

[29] **C. Gunnar** and, **F. X. Campion**, Machine Intelligence for HealthcareCreateSpace Independent Publishing

[30] **A. Agah**, Medical Applications of Artificial Intelligence. Boca Raton, FL: CRC Press, 2017.

[31] **S. M. Richins**, Emerging Technologies in Healthcare. Boca Raton, FL: CRC Press, 2015.

[32] **S. Russell** and **P. Norvig**, Artificial Intelligence: A Modern Approach. Upper Saddle River, NJ: Prentice Hall, 3rd edition, 2009.

[33] **P. S. Mahajan**, Artificial Intelligence in Healthcare. Parag Suresh Mahajan, 2018.

[34] **A. Bohr** and K. **Memarzadeh** (eds.), Artificial Intelligence in Healthcare. Academic Press, 2020.

CHAPTER 5
AI IN ENGINEERNING

"If an AI possessed any one of these skills—social abilities, technological development, economic ability—at a superhuman level, it is quite likely that it would quickly come to dominate our world in one way or another. And as we've seen, if it ever developed these abilities to the human level, then it would likely soon develop them to a superhuman level. So, we can assume that if even one of these skills gets programmed into a computer, then our world will come to be dominated by AIs or AI-empowered humans." — *Stuart Armstrong*

5.1 INTRODUCTION

For decades researchers have been fascinated by the notion of creating a machine that can replicate the human brain. Such a machine is known as artificial intelligence. Artificial intelligence (AI) refers to systems that act intelligently, whether in a specific domain (narrow AI), or in general (strong AI). It is a computer system's ability to mimic human behavior, as shown in Figure 5.1 [1].

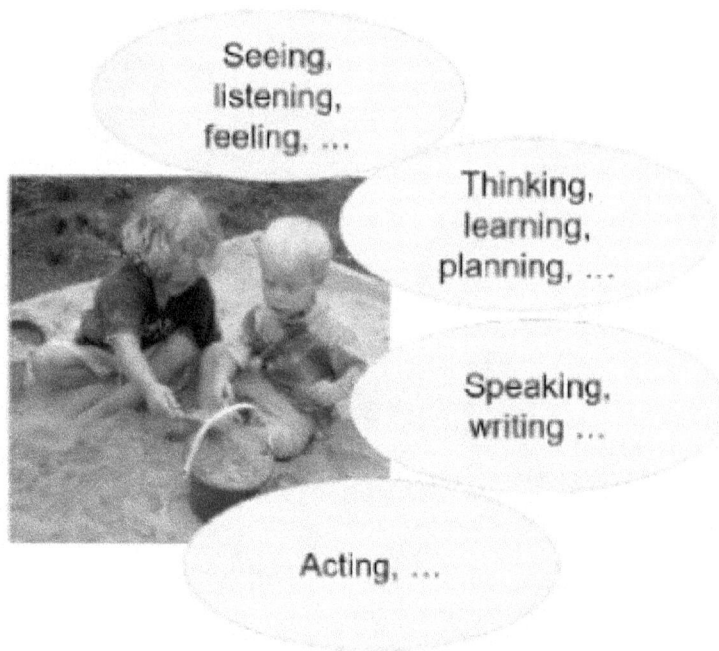

Figure 5.1 Behavior of human intelligence [1].

It is a system that thinks like humans and acts like humans. Machines can perform human-like tasks and demonstrate intelligence that is comparable to natural intelligence that humans and animals demonstrate.

Artificial intelligence has endless potential to handle tasks commonly done by humans, including natural language processing, image recognition and data analytics, visual perception, decision-making, speech recognition, business process management, and even the diagnosis of disease, all of which normally require human intelligence. Artificial intelligence is now everywhere and has a great deal to offer the world of engineering. As shown in Figure 5.2, AI has had an impact on just about every field, including engineering.

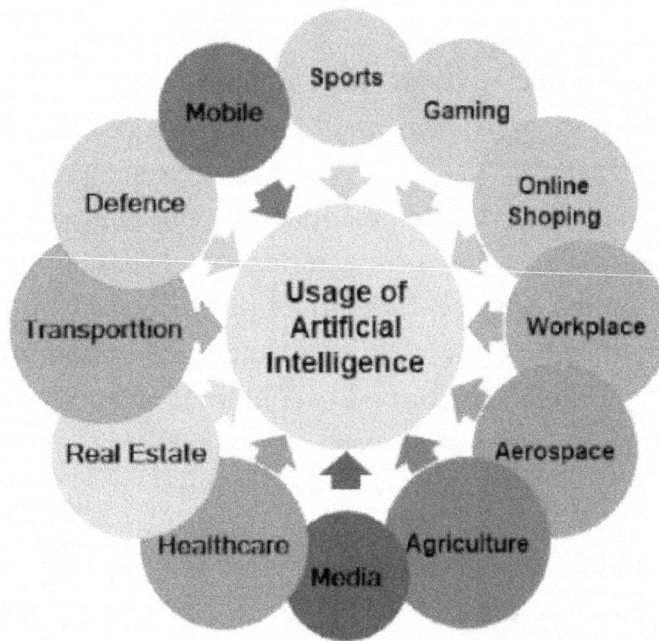

Figure 5.2 Industries impacted by AI revolution.

Some of the most interesting applications of artificial intelligence are in the field of engineering. For this reason, more and more people are becoming AI Engineers [2]. The chapter provides an overview of the advances of artificial intelligence applied in engineering. It begins with a brief review on AI and its related field of artificial intelligence engineering. It discusses what AI engineers do. It highlights the benefits and challenges of application of AI in engineering. The last section concludes with comments.

5.2 REVIEW ON ARTIFICIAL INTELLIGENCE

Artificial intelligence (AI) describes computer systems and computer software that are capable of intelligent behavior. It is a machine-exhibited intelligence rather than humans based. The primary goal of AI is to explore how to imitate and execute some of the intelligent functions of the human brain. AI is becoming an essential part of our lives and we are consuming its services consciously or unconsciously. Today, AI is the most important general-purpose technology. It is applied in many fields such as expert system, knowledge base system, intelligent database system, and intelligent robot system, healthcare, transportation, business, finance, and engineering [3].

The term "artificial intelligence" (AI) was first used at a Dartmouth College conference in 1956. AI is now one of the most important global issues of the 21st century. AI is the branch of computer science that deals with designing intelligent computer systems that mimic human intelligence (e.g., visual perception, speech recognition, decision-making, and language translation). The ability of machines to process natural language, to learn, and to plan makes it possible for new tasks to be performed by intelligent systems. The main purpose of AI is to mimic the cognitive function of human beings and perform activities that would typically be performed by a human being. Without being taught by humans, machines use their own experience to solve a problem.

AI is not a single technology but a range of computational models and algorithms. The major disciplines in AI include expert systems, fuzzy logic, and artificial neural networks (ANNs), machine learning, deep learning, natural language processing, computer vision, and robotics. These tools or technologies are illustrated in Figure 5.3.

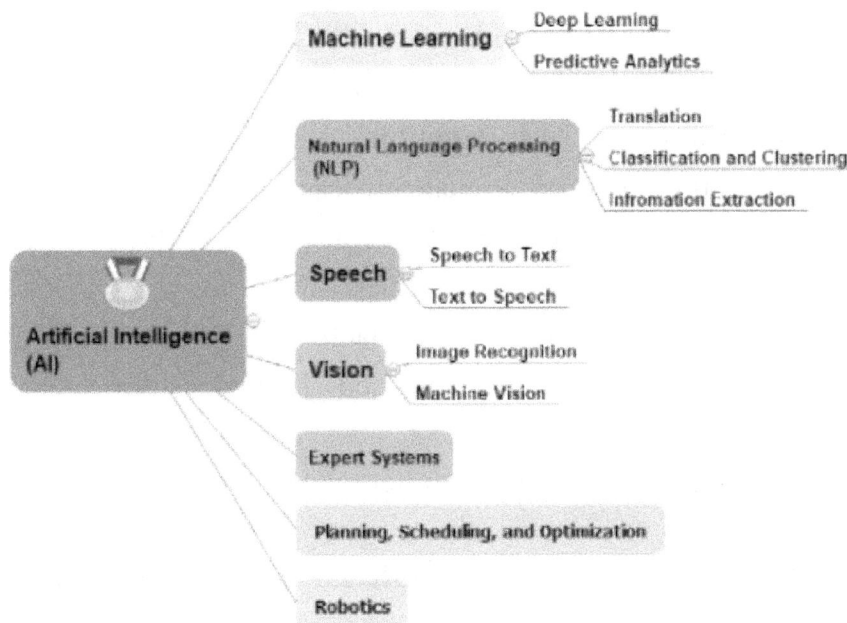

Figure 5.3 Many branches of artificial intelligence.

They have been used to achieve AI's goals. It has been observed that a combination of AI techniques performed better than when each one of the AI techniques is used singly. The five AI tools that are most applicable to engineering problems are knowledge-based systems, fuzzy logic, inductive learning, neural networks and genetic algorithms [4].

Ever since the field of AI was founded in 1956, it has made explosive, surprising advances. This is evident by the development of AlphaGo, autonomous cars, Alexa, Watson, game playing, robotics, computer vision, speech recognition, and natural language processing. Historically, the term AI is reflected collectively to the following branches [5]:

- Game playing—for example, Chess, Go Symbolic reasoning and theorem-proving—for example, Logic Theorist, MACSYMA
- Robotics—for example, self-driving cars
- Vision—for example, facial recognition
- Speech recognition, Natural language processing–for example, Siri
- Distributed & evolutionary AI—for example, drone swarms, agent-based models
- Hardware for AI—for example, Lisp machines
- Expert systems or knowledge-based systems—for example, MYCIN, CONPHYDE
- ML—for example, clustering, deep neural nets, Bayesian belief nets.

5.3 ARTIFICIAL INTELLIGENCE ENGINEERING

AI engineering is a relatively new field. It is essentially the use of algorithms, computer programming, neural networks, and other related technologies in the development of AI applications and techniques. The field aims to equip practitioners to ensure human needs are translated into understandable, ethical, and trustworthy AI. It is a field of research and practice that combines systems engineering, software engineering, computer science, and human-centered design to create AI systems in accordance with human needs. AI has produced a number of powerful tools to solve difficult problems normally requiring human intelligence [6,7]

Five ways in which AI may impact the engineering profession [8]:

1. History has shown that technological advances in the past have helped create new jobs. This will be especially relevant for those in the engineering community.

2. With the rapid evolution of technology, there will be an increased need for engineers to research, create, and test AI systems.

3. Engineers have an enormous opportunity to showcase their creativity in response to advances in AI.

4. New types of experts will increasingly be in demand in response to the new types of work created by AI technology.

5. New developments in AI will enable engineers to complete their work more efficiently and solve a wide range of problems.

To pursue an AI engineering role within your organization requires relevant knowledge of the business in addition to any necessary technical skills. To be relevant and create true value for the organization, artificial intelligence engineering must go beyond the technical to provide applications specific to the business. AI-centric organizations are creating jobs for AI engineers and attracting people who can perform a hybrid of data engineering, data science, and software development tasks. Sometimes the software engineers of an organization can quickly navigate and evolve into AI engineers.

5.4 AI ENGINEERS

A bachelor's degree is usually required to become an artificial intelligence engineer. AI engineers are responsible for the design, implementation, and management of AI-based tools throughout the organization. They must have a sound understanding of programming, software engineering, and data science. To become well-versed in AI, it is important that AI engineers learn programming languages, such as Python, R, Java , and C++ to build and implement models. R is widely used programming languages in artificial intelligence.

AI engineers are in high demand because designing intelligent systems is not easy task. They are needed to provide the expertise a company needs to build models. Typically, AI engineers design, build, maintain, and deploy AI-based systems. AI engineers design machines that are capable of replicating just about anything that humans can do. They are primarily responsible for implementing machine learning. Depending on the industry, the responsibilities of AI engineers include the following [9,10]:

- Analyze data, design software, create AI algorithms, perform image processing, and implement natural language processing.

- Convert the machine learning models into application program interfaces (APIs) so that other applications can use it.

- Build AI models from scratch and help the different components of the organization (such as product managers and stakeholders) understand what results they gain from the model.

- Build data ingestion and data transformation infrastructure
- Automate infrastructure that the data science team uses
- Perform statistical analysis and tune the results so that the organization can make better-informed decisions.

- Set up and manage AI development and product infrastructure
- Be a good team player, as coordinating with others is a must
- Have excellent problem-solving skills to resolve obstacles for decision making
- Need to communicate well about their products and ideas to stakeholders. An AI engineer needs to be able to speak the language of various groups he will work within the organization.

AI engineers find job positions in several areas such as enterprises, businesses, healthcare, manufacturing, finance, and fraud detection. Examples of companies that hire AI engineers include Amazon, Accenture, IBM, NVIDIA, Microsoft, Intel, Facebook, Lenovo, Samsung, and Adobe. Countries hiring most AI engineers include US, China, UK, Canada, Germany, Russia, and India [11]. Figure 5.4 shows some AI engineers at work [12].

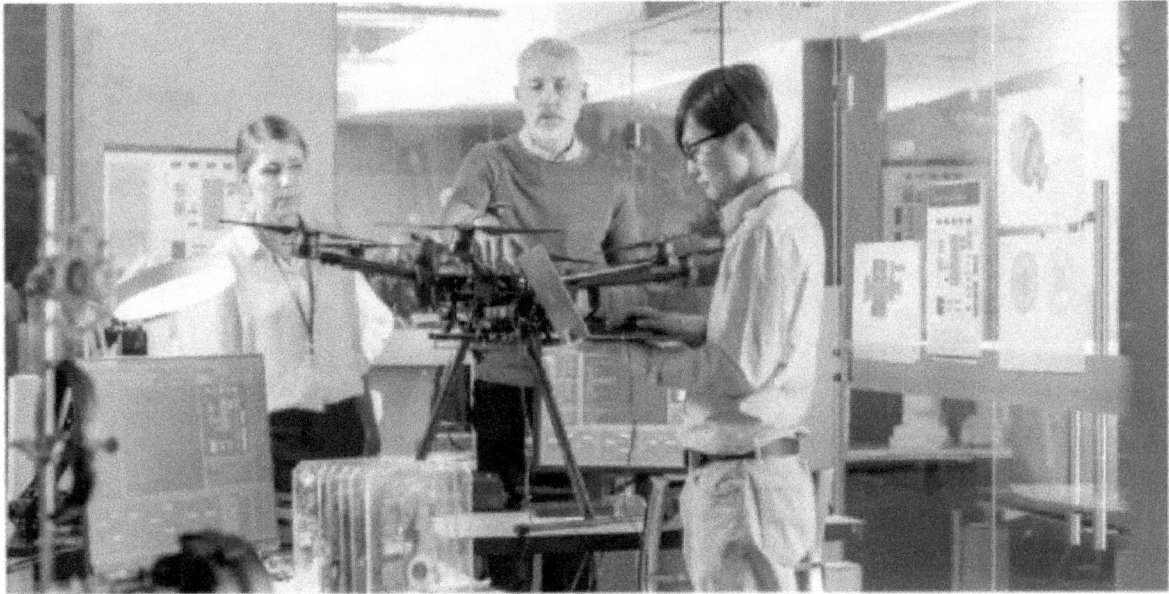

Figure 5.4 Typical AI engineers at work [12].

5.5 APPLICATIONS OF AI IN ENGINEERING

AI engineering has matured to the point in offering real practical benefits in many of their applications. With advances in technology, the application of AI engineering is becoming more and more extensive. There are a number of applications of AI that are of considerable use to engineers, especially in the industrial and manufacturing contexts. The AIs used in the engineering sector combine both software and hardware components. They are used in the following areas [13].

- **Manufacturing:** One popular application of AI in engineering is in the field of automobile manufacturing. AI is increasingly finding its way into manufacturing sector. It is indisputable that AI plays a critical role in the future of manufacturing engineering. From maintenance to virtual design, AI enables manufacturing of modern, advanced, and custom products. Robots were originally capable of performing simple engineering tasks that involve large movements. These machines can now perform precision work, emulate intricate tasks, and handle complicated manufacturing tasks. In spite of the anticipated unemployment fears due to the adoption of AI, the manufacturing industry will experience higher production and higher efficacy. Figure 5.5 shows a workplace automation and manufacturing [10].

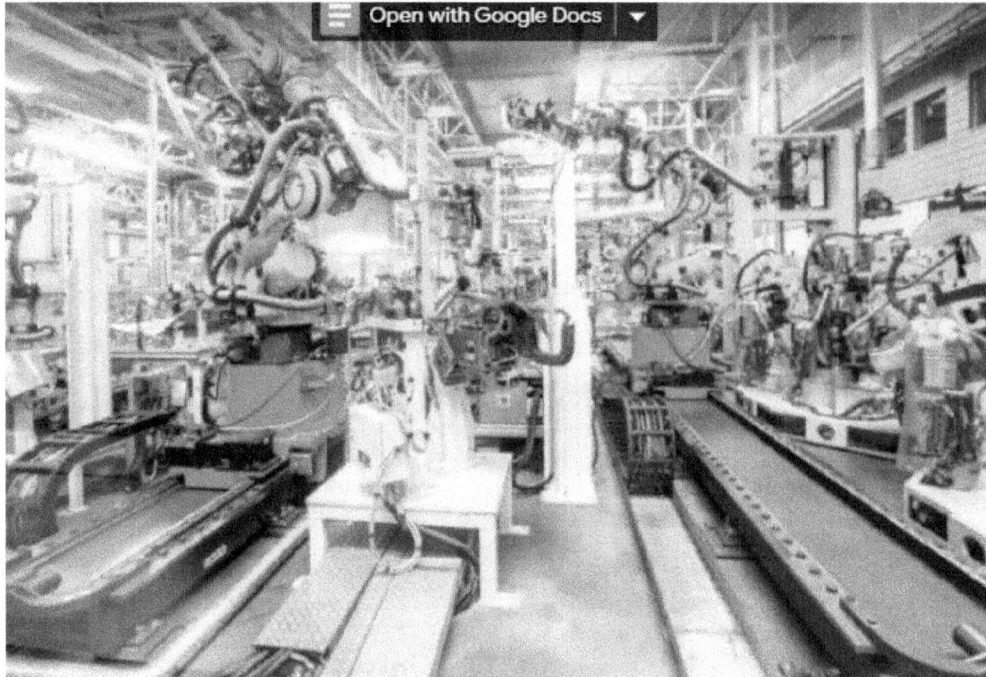

Figure 5.5 Workplace automation [10].

- **Autonomous Vehicles:** One of the main changes in AI engineering is the automation of many low-level engineering tasks. Engineers build systems of connected sensors and cameras that ensure that an autonomous vehicle's AI can "see" the environment. Autonomous vehicles include cars, trains, submarines, aircrafts, ships, etc. For instance, a self-driving car can navigate impressively through traffic. With companies like Tesla, Google, and Uber already testing self-driving vehicles on the road, many of the biggest remaining barriers are legislative rather than technological. Figure 5.6 is a demonstration of a Tesla Model self- driving without any human intervention [14].

Figure 5.6 Demonstration of a Tesla Model self - driving without any human intervention [14].

- **Big Data:** Data is everywhere. Many processes and functions in engineering requires the usage and storage of data. Most engineers work on big data. Big data analysis can reveal some truth to engineers and help them make decisions that are based on objective, scientific analyses. AI engineers will require some knowledge of big data technologies such as Apache Spark, Hadoop, Cassandra, and MongoDB.

- **Business:** AI engineers can quickly add machine learning capabilities to business-critical systems such as enterprise resource planning (ERP), customer relationship management (CRM), and mobile device management (MDM). Artificial intelligence engineering is enabling organizations to create hybrid operating environments that combine data science, data engineering, and software development. Successful AI projects will deliver value to the enterprise and address the relevant pain points of the business [15].

- **Signal Processing:** The adoption of machine learning in engineering has been especially valuable for signal processing. Signal processing techniques can also be used to improve the data fed into machine learning systems. By cutting out much of the noise that would otherwise be included in these inputs, engineers achieve cleaner results in the performance of Internet-of-things devices and other AI-enabled systems. Life-changing possibilities that can come from applying AI to investigations in signal processing [5].

- **Civil Engineering:** AI-based solutions can provide valuable means for efficiently solving problems in the civil engineering. Civil engineers are expected to deliver practical sustainable solutions for engineering projects. AI tools used in civil engineering includes evolutionary computation, neural networks, fuzzy systems, expert system, reasoning, classification, and learning. AI technology is widely used in many areas for civil engineering field, such as civil building engineering, construction engineering, underground engineering, bridge engineering, road engineering, geological exploration, geotechnical engineering, construction management, hydraulic optimization, transportation engineering disaster prevention project, material engineering, petroleum chemical industry, etc. [16].

- **Electrical Engineering:** As electrical engineers work at the forefront of technological innovation; their pursuits often overlap with artificial intelligence. For decades, electrical engineers have explored how different AI tools can be applied to electrical systems. Machine learning and electrical engineering professionals leverage AI to build and optimize systems. Implementing the latest engineering applications of AI may prove to be a path for career advancement on the cutting edge of the field. The most promising achievements at the intersection of AI and electrical engineering have focused on autonomous vehicles, power systems, signal processing, smart grid, and smart energy [17].

- **Chemical Engineering:** Many issues in chemical engineering can benefit from AI techniques. Major efforts to developing AI tools for chemical engineering problems started in the early 1980s. Today, chemical engineering is at an important crossroads and is undergoing an unprecedented transition. Many chemical engineers are excited about the potential applications of AI. They realize that AI will play an increasingly dominant role in chemical engineering research and education. Chemical engineers have always prided themselves on their modeling capabilities, but, in this new era, modeling would go beyond the differential and algebraic equations [DAEs]. AI will play a major role in chemical engineering in the coming years [18].

- **Mechanical Engineering:** This is a branch of engineering dealing with design, construction, and machines. Mechanical engineering plays a catalyst in imagining a healthy, secure, connected, and creative future ahead. The field is constantly upgrading from traditional mechanical engineering to the mechatronics engineering and AI. Within this field, mechanical design or engineering works is the area that mostly applies artificial intelligence. AI technology is often used in the diagnosis of mechanical engineering failure. Mechanical design is a process to design the required component, system, or process. It embraces product design, machine design, mechanical component design, tooling and fixture development, mold design, and casting design. Other areas of AI impact include manufacturing, production, product characteristics, big data storage, stress estimation of 3D structures, material evaluation for different services, and structure generation [19]. Figure 5.7 shows artificial intelligence in mechanical engineering [20].

Figure 5.7 Artificial intelligence in mechanical engineering [20].

- **Software Engineering:** This is a knowledge-intensive activity that cuts across all the stages of the software development phases. It presumably requires intelligence, and it seems natural to use AI techniques to build systems. In fact, AI engineering may be regarded as primarily an extension of software engineering. It is customary to divide software engineering activities into two categories:

programming-in-the-small and programming-in-the-large. Programming-in-the-small is usually done by individuals or small groups. Programming-in-the-large is done by large groups of people resulting in programs hundred thousand or millions of lines [16b]. AI techniques like artificial neural networks, fuzzy logic, machine learning and data mining have been used for solving several software engineering problems.

They have continued to have a considerable impact on computer software development. Software products become pervasive in all areas of society. AI is potentially a game changer that improves software quality. Applications of AI in software engineering form a new class of tool support called software analytics [21]. AI provides robust approaches for software development in order to evaluate complex software. AI has been used to reduce human efforts in software engineering activities.

- **Systems Engineering:** This is an interdisciplinary field of engineering development and engineering management that focuses on how to design and manage complex systems. Activities necessary for the successful development, design, and implementation become more difficult when dealing with large or complex projects. AI is becoming more and more relevant to systems engineering [22].

- **Transportation Engineering:** Transportation problems exhibit a number of characteristics that make them amenable to solution using AI techniques. Transportation engineers are expected to meet the goals of providing safe, efficient, and reliable transportation while minimizing the impact on the environment. To achieve this goal, transportation engineers along other transportation professionals have explored the feasibility of applying AI tools to address real transportation problems and improve the efficiency, safety, and environmental-compatibility of transportation systems [23].

- **Machine Design:** To design is to "devise courses of action aimed at changing Existing situations into preferred ones." Much of the technical aspect of engineering will be moved to the machine-based design system. Advances in AI and other technologies such as cognitive computing, Internet of Things, 3D (or even 4D) printing, advanced robotics, virtual and mixed reality, and human–machine interfaces are transforming what, where, and how products are designed, manufactured, assembled, distributed, serviced, and upgraded [24].

Other applications of AI in engineering include engineering education, concurrent engineering, electronic engineering, geotechnical engineering, engineering knowledge, petroleum engineering, tissue engineering, optical engineering, structural engineering, product engineering construction engineering, traffic engineering, process engineering, safety engineering, aerospace engineering, medical engineering, biomedical engineering, military engineering, informatics, pattern recognition, contextual advertising, language translation, visual identification or perception, fault diagnosis, image processing, microscopic defect identification, oil and gas, and risk management.

5.6 BENEFITS

There is no denying that AI engineering has been successful. There are different areas where AI impacts the engineering process. Even though these successes may be in narrow domains, they have caused massive disruptions. Compared to human speed, AI is much faster. AI will cause some positive big changes to the engineering profession. When combined with data analysis, AI and ML facilitate predictive analytics. A knowledge of deep learning provides AI engineers with the skills necessary to build AI models with unstructured data. AI engineers can design systems optimized to perform machine learning tasks efficiency. This opens the door for new possibilities in autonomous vehicle guidance, customer relationship management, fraud detection, and several other applications.

5.7 CHALLENGES

In many places, there are widespread fears and anxieties surrounding the impact of automation on jobs. It is estimated that half of the jobs in the United States are at risk of being automated over the next decade or two. AI is poised to replace people in certain kinds of jobs, such as in the driving of vehicles and repetitive manufacturing tasks. It can be the source of disruptions as new types of tasks are created and other kinds of work become less needed due to automation. This may dramatically affect the job descriptions of engineers [25]. AI technology poses questions for both civil and criminal law. Large-scale deployment of AI-driven solutions could have both positive and negative impacts on the environment. Researchers tend to agree that the long-term benefits of automation outweigh the potential drawbacks. It is not likely that engineers will be replaced by the machines they created.

5.8 CONCLUSION

The term "artificial intelligence" refers to a variety of systems built to imitate how a human mind makes decisions and solves problems. These are promising times for artificial intelligence engineering. Artificial intelligence is gradually changing our world. Today, expectations for AI are sky high. AI is expected to be the most disruptive technology category for the next decade due to the advances in related areas. AI will significantly increase the efficiency of the existing economy and transform the global economy.

With more and more organizations relying on AI tasks as part of their daily operation, demand for AI engineers to work with AI-related technologies will only increase AI is poised to create many exciting job opportunities in the coming years. AI engineering as an academic program is being offered by some universities at the undergraduate and graduate levels, providing graduates who can provide the desired mix of programming experience, mathematical knowledge, and statistical skills. Young professionals can benefit immensely from entering this burgeoning field. It takes initiative, determination, perseverance, and know-how to be an AI engineer. More information on artificial intelligence in engineering can be found in the books in [1,26-44] and following related journals:

- Artificial Intelligence Review
- Artificial Intelligence in Engineering
- Engineering Applications of Artificial Intelligence
- Journal of Healthcare Engineering

REFERENCES

[1] **B. G. Humm**, Applied artificial intelligence: An Engineering Approach. Lean Publishing, 2016.

[2] "24 industries to be disrupted by AI: An infographic," https://www.oneragtime.com/24-industries-disrupted-by-ai-infographic/
[3] https://www.researchgate.net/figure/Fields-of-artificial-intelligence-10_fig1_324183626

[4] **D.T. Pham** and **P. T. N**. Pham, "Artificial intelligence in engineering," International Journal of Machine Tools and Manufacture, vol. 39, no. 6, June 1999, pp. 937-949.

[5] **V. Venkatasubramanian**, "The promise of artificial intelligence in chemical engineering: Is it here, finally?" AIChE Journal, vol. 65, no. 2, February 2019, pp. 466-478.

[6] **D.T. Pham** and **P.T.N. Pham**, "Artificial intelligence in engineering," International Journal of Machine Tools & Manufacture, vol. 39, 1999, pp. 937–949.

[7] **M. N. O. Sadiku**, **T. J. Ashaolu**, and **S. M. Musa**, "Artificial intelligence in medicine: A primer," International Journal of Trend in Research and Development, vol. 6, no. 1, January-Feb. 2019, pp. 270-272.

[8] "The impact of artificial intelligence on the engineering profession," https://iconnectengineers.com/blog/impact-artificial-intelligence-engineering-profession/

[9] **J. Hughes**, "The key roles of AI engineers," October 2019, https://engineeringmanagementinstitute.org/key-roles-ai-engineers/#:~:text=AI%20engineers%20build%2C%20maintain%20and,and%20service%20delivery%2C%20among%20others.

[10] "How to become an artificial intelligence engineer?" December 2020, https://www.simplilearn.com/tutorials/artificial-intelligence-tutorial/how-to-become-an-ai-engineer

[11] "The artificial intelligence engineer career roadmap - All you need to know" September 2019, https://www.artiba.org/blog/the-artificial-intelligence-engineer-career-roadmap-all-you-need-to-know

[12] **R. Johnson**, "Jobs of the future: Starting a career in artificial intelligence," May 2020, https://www.bestcolleges.com/blog/future-proof-industries-artificial-intelligence/

[13] **M. N. O. Sadiku**, **U. C. Chukwu**, **A. Ajayi-Majebi**, and **S. M. Musa**, "Artificial intelligence in engineering," Journal of Scientific and Engineering Research, vol. 7, no. 12, 2020, pp. 185-193.

[14] **M. Alba**, "Artificial intelligence and engineering," April 2017, https://www.engineering.com/DesignerEdge/DesignerEdgeArticles/ArticleID/14723/Artificial-Intelligence-and-Engineering.aspx

[15] **T. Brown**, "What is artificial intelligence engineering? Prospects, opportunities, and career outlooks," May 2020, https://itchronicles.com/artificial-intelligence/what-is-artificial-intelligence-engineering-prospects-opportunities-and-career-outlooks/

[16] **P. Lu**, **S. Chen**, and **Y. Zheng**, "Artificial intelligence in civil engineering," Mathematical Problems in Engineering, 2012.

[17] "Applications of AI and machine learning in electrical and computer engineering," July 2020,
https://online.egr.msu.edu/articles/ai-machine-learning-electrical-computer-engineering-applications/

[18] **R. Johnson**, "Jobs of the future: Starting a career in artificial intelligence," May 2020,
https://www.bestcolleges.com/blog/future-proof-industries-artificial-intelligence/

[19] "Artificial intelligence and its alliance with mechanical engineering," February 2020,
https://www.teslaoutsourcingservices.com/blog/artificial-intelligence-and-its-alliance-with-mechanical-engineering/

[20] Artificial intelligence and its alliance with mechanical engineering," February 2020,
https://www.teslaoutsourcingservices.com/blog/artificial-intelligence-and-its-alliance-with-mechanical-engineering/

[21] **D. Barstow**, "Chapter 16 - Artificial intelligence and software engineering," Exploring Artificial Intelligence: Survey Talks from the National Conferences on Artificial Intelligence, 1988, pp. 641-670.

[22] **H. K. Dam**, "Artificial intelligence for software engineering," XRDS, vol. 25, no. 3, Spring 2019, pp. 34-37.

[23] "Artificial Intelligence for Systems Engineering - AI4SE 2020," October 2020
https://www.incose.org/events-and-news/search-events/2020/10/13/default-calendar/artificial-intelligence-for-systems-engineering---ai4se-2020

[24] "Artificial intelligence in transportation," Transportation Research, No. E-C113 January 2007.

[25] **A. K. Noor**, "AI and the future of the machine design," Mechanical Engineering, vol. 139, no. 10, October 2017, pp. 38-43.

[26] **G. Rzevski** and **R. A. Adey** (eds.), Applications of Artificial Intelligence in Engineering VI. Springer, 1991.

[27] **Z. Shi**, Advanced Artificial Intelligence. World Scientific, 2nd edition, 2020.

[28] **R. J. Schalkoff**, Artificial Intelligence: An Engineering. McGraw-Hill Education, 1990.

[29] **C Tong** and **D Sriram** (eds.), Artificial Intelligence in Engineering Design: Volume III: Knowledge Acquisition, Commercial Systems, And Integrated Environments. Academic Press, 2012.

[30] **T. E. Quantrille** and **Y. A. Liu,** Artificial Intelligence in Chemical Engineering. Academic Press, 2012.

[31] **C. Rich** and R. **C. Waters,** Readings in Artificial Intelligence and Software Engineering. Morgan Kaufmann, 2014.

[32] **T. W. Rondeau** and **C. W. Bostian**. Artificial Intelligence in Wireless Communications. Boston, MA: Artech House, 2009.

[33] **F. Meziane** and **S. Vadera**, Artificial Intelligence in Software Engineering: Current Developments and Future Prospects. IGI Global, 2010

[34] **A. K. Luhach** and Atilla **Elçi**, Artificial Intelligence Paradigms for Smart Cyber-Physical Systems. IGI Global, 2021.

[35] **I. Smith (ed.)**, Artificial Intelligence in Structural Engineering: Information Technology for Design, Collaboration, Maintenance, and Monitoring. Springer, 1998.

[36] **G. Bekda**, S. **M. Nigdeli**, and M. **Yücel (eds.)**, Artificial Intelligence and Machine Learning Applications in Civil, Mechanical, and Industrial Engineering (Advances in Computational Intelligence and Robotics). IGI Global, 2019.

[37] L. **P. Suresh, S. S. Dash**, and **B. K. Panigrahi (eds.)**, Artificial Intelligence and Evolutionary Algorithms in Engineering Systems: Proceedings of ICAEES 2014, Volume 1. Springer 2015.

[38] **M. Bielli, G, Ambrosino**, and **M. Boero (eds.)**, Artificial Intelligence Applications to Traffic Engineering. Taylor & Francis, 1994.

[39] **C. Tong** and D. **Sriram (eds.)**, Artificial Intelligence in Engineering Design: Volume III: Knowledge Acquisition, Commercial Systems, And Integrated Environments. San Diego, CA: Academic Press, 2012.

[40] **R. J. Schalkoff,** Artificial Intelligence: An Engineering Approach. McGraw-Hill, 1990.

[41] **T. E. Quantrille** and **Y. A.** Liu, Artificial intelligence in chemical engineering. San Diego, CA: Academic Press, 2012.

[42] **C. Rich** and **R. C. Waters (eds.)**, Readings in Artificial Intelligence and Software Engineering. Elsevier, 2009.

[44] **F. Meziane**, and S. **Vadera**, Artificial Intelligence Applications for Improved Software Engineering Development: New Prospects. Information Science Reference, 2009.

CHAPTER 6
AI IN EDUCATION

"Teachers will not be replaced by technology, but teachers who do not use technology will be replaced by those who do." —*Hari Krishna Arya*

6.1 INTRODUCTION

According to Nelson Mandela, "Education is the most powerful weapon which you can use to change the world." Education is a process where teachers give systematic instructions, while students receive them. It is a major determining factor for an individual's success in life. Education seems to be fixed in terms of time, place, and prescribed activities. Learning takes place continuously, especially for younger people. Traditional education systems are known to be inflexible but is now changing to adapt to the technological advancements of today's world. One key technology that is poised to transform education is artificial intelligence (AI). The implementation of AI has several benefits for students and teachers alike [1].

AI is a general-purpose technology that can perform tasks that previously required human beings. It can be regarded as the simulation of human mental capacity by the machines. It is a domain of research with many sub-disciplines, domains of expertise, and developmental dynamics [2]. As shown in Figure 6.1, AI is rapidly transforming many industries such as healthcare, transportation, business, retail, construction, food industry, manufacturing, and education. Apple's Siri voice assistant, Amazon's shopping recommendations, Uber ride sharing, and Google translate are common examples of how AI has invaded our daily lives [3].

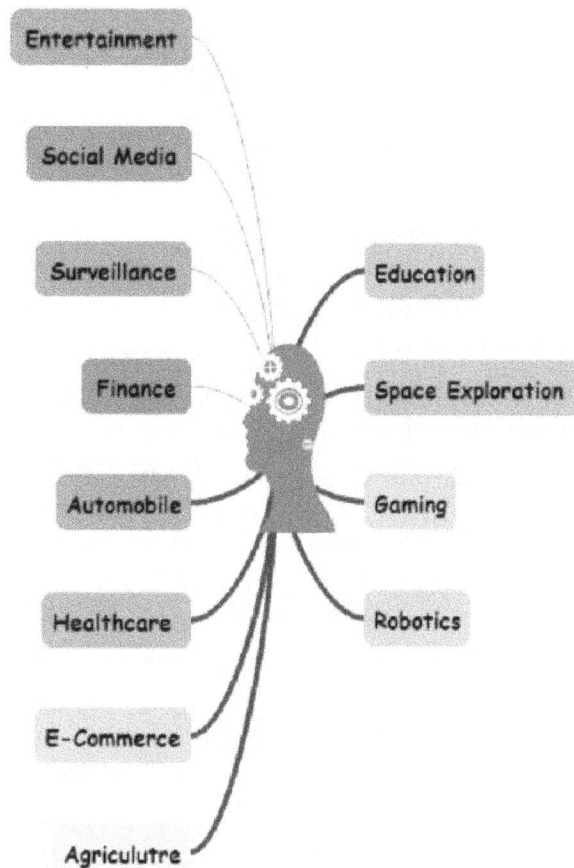

Figure 6.1 Industries impacted by AI revolution.

AI has impacted almost all sectors of human life and provided a lot of benefits to various fields, including education. It is having a positive impact on educational applications. The field of education is a place where artificial intelligence is poised to make big changes. Artificial intelligence in education (AIED) has become a field of scientific research for more than 30 years. AIED is mainly concerned with the development of "computers which perform cognitive tasks, usually associated with human mind" [4].

This chapter provides various applications of AI in education. It begins by giving a brief review of AI and various possibilities of AI in education. It covers how AI can be applied in elementary school, high school, and higher institutions. It presents some applications of AI in education. It discusses global adoption of AI in education. It highlights the benefits and challenges of AI in education. It considers the future of education as far as AI is concerned. The last section concludes with comments.

6.2 REVIEW ON ARTIFICIAL INTELLIGENCE

Artificial intelligence refers to the ability of a computer system to perform human tasks (such as thinking and learning) that usually can only be accomplished using human intelligence. AI technology in education is allowing a degree of flexibility and customization that was never before possible. It is revolutionizing schools and classrooms, making educator's job a lot easier. It is poised to revolutionize education. AI is not a single technology but a range of computational models and algorithms. The major disciplines in AI include expert systems, fuzzy logic, artificial neural networks (ANNs), machine learning, deep learning, natural language processing, computer vision, and robotics. These computer-based tools or technologies have been used to achieve AI's goals are illustrated in Figure 6.2.

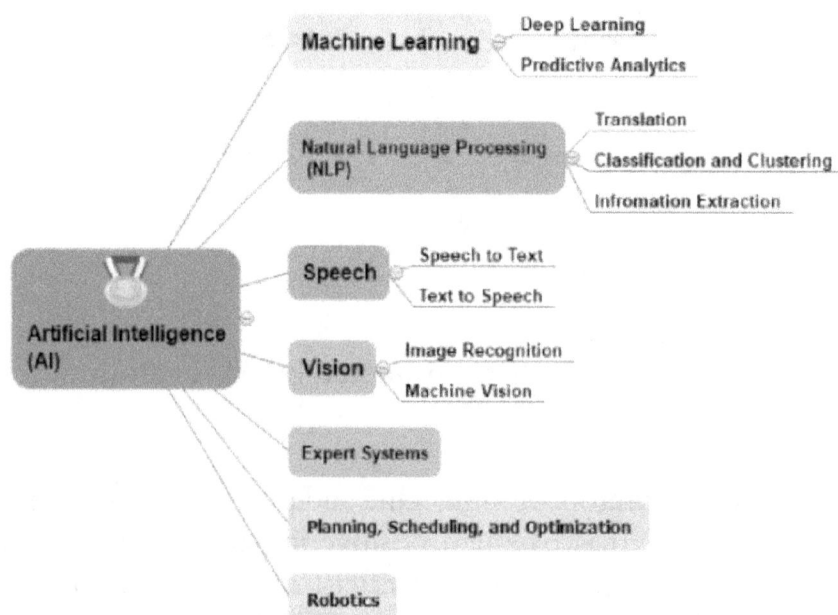

Figure 6.2 Many branches of artificial intelligence.

Each AI tool has its own advantages. Using a combination of these models, rather than a single model, is recommended.

6.3 POSSIBILITIES FOR AI IN EDUCATION

Intelligent tutoring system (ITS), intelligent learning system (ILS) technologies, virtual reality (VR), and augmented reality (AR) education technology use artificial intelligence to create more learning environments. Advances in these technologies are fueling the applications of AI in education, which include training, communications, administration, and resource management [7].

Educators may soon find themselves teaching in a digital classroom with AI at its heart. The possibilities for artificial intelligence in education include the following [8]:

1. **Communication**: The pleasure of live communication is crucial for creating a favorable emotional atmosphere in a class. Students and teachers will be able to communicate instantly with one another as well as to connect with other forms of AI around the world. Information from anywhere in the world is at students' fingertips.

2. **Personalization**: Unlike the traditional education system, AI-based systems promise to serve students of all capabilities and discover their full potential. Personalized learning opportunities can be offered for students. Students now have the advantage of personalized study plans, accessible from any user device or location. AI-based tools are helping them organize classes and customize the experience to match the student's ethnicity.

3. **Assessments**: The formulation of assessment instruments using AI techniques can provide more conducive and diagnostic outcomes than what conventional tests. can yield. AI could help teachers to assess students and streamline the grading process, with the added benefit of being able to quickly take the data, provide an analysis for teachers, so that time can be saved for more classroom interactions. Assessing students' motivation, attention, and engagement could enhance their learning.

4. **Intelligent Tutoring Systems (ITS)**: ITS has become popular since the early 2000s. It is essentially a computer system that aims to provide immediate and customized instruction or feedback to learners, usually without requiring intervention from a human teacher. ITS can be used to simulate one-to-one personal tutoring. While far from the norm, they are capable of functioning without a teacher having to be present and can effectively challenge and support the learner using different algorithms.

5. **Virtual Reality Learning**: Virtual reality (VR) is a simulated experience that enhances learning and engagement by allowing user to view and interact with virtual items. It provides users with immersive experience via three-dimensional (3-D) visual and auditory simulations. It brings vivid experiences to the classroom for the purpose of increasing student engagement and enhancing learning. Combined with virtual and augmented reality, AI can bring a dynamic, immersive learning environment to the classroom. VR-assisted learning allows for educational support in authentic environments and extends the boundaries of the classroom. This will also act as a steppingstone to real-world experiences, with fully integrated AIs.

The future likely holds a lot of possibilities for AI and teachers who can take the opportunity to be informed of the possibilities and being open to discussions with students.

6.4 AI IN DIFFERENT LEVELS OF AI

Applications of AI in education are on the rise and are receiving a lot of attention at all levels of education. The promise of AI applications lies partly in their efficiency and efficacy. School children, and even kindergartners, who live in this generation are regarded as digital natives. Digital natives are children who were born after 1980. Early exposure to technology like the Internet, computers, and mobile devices fundamentally changes the way digital natives learn, behave, and operate [9]. Educators serve as the link to preparing students to thrive in a world where AI is an integral part of their life and career.

- **Elementary School**: School children have grown up with technology at their fingertips, in a world where education has changed, and where the Internet is their primary source of entertainment and information. Chalkboards in the classroom is also a thing of the past. The availability of smart technology such as interactive whiteboards, IT suites, and tablet-based learning are becoming more commonplace in elementary schools. Nao is a humanoid robot that talks, moves, and teaches children from ages seven and up everything from literacy to computer programming. This way, students are prepared to work alongside AI and develop the skills required for them to thrive in a digital workplace [10].

- **High School**: AI can be integrated in K-12 education. AI should be a critical element of any STEM curriculum. There is no rulebook for deploying AI in schools. Querium uses AI to deliver customizable STEM tutoring lessons to high school and college students. High school students with exposure to AI tool will be well prepared for making decisions involving AI technology, regardless of what industry they find themselves. Therefore, it is of great importance to familiarize young people in school with the technical background and the underlying AI concepts [11].

- **Higher Education**: The AI technology will soon have a significant impact on higher education institutions. It is linked to the future of higher education. Some faculty members, teaching assistants, student counsellors, and administrative staff may fear that intelligent tutors, expert systems and chat bots will take their jobs. Institutions of higher education increasingly rely on algorithms for marketing to prospective students, enrollment, planning curricula, and allocating resources such as financial aid and facilities. Some AI applications provide student guidance and help students automatically schedule their course load. Colleges and universities can use AI for instruction by using educational software [12].

6.5 APPLICATIONS OF AI IN EDUCATION

Figure 6.3 shows a typical example on how AI is changing the education industry [13].

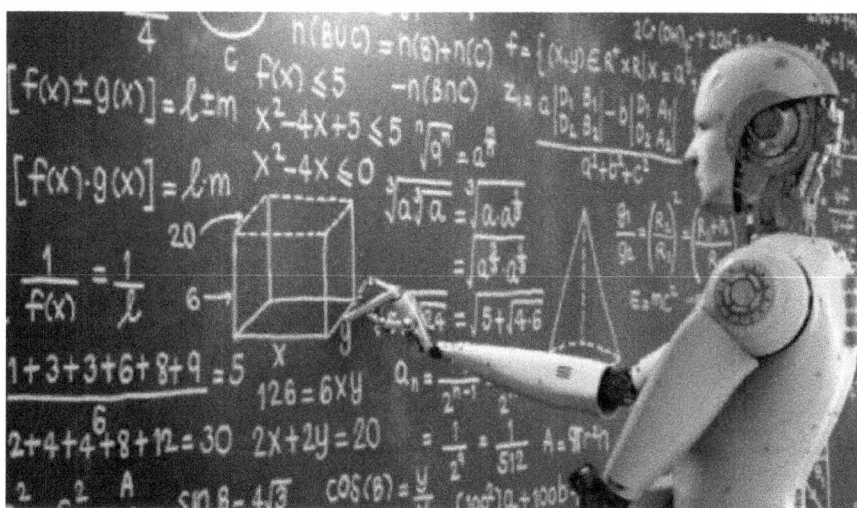

Figure 6.3 A typical example on how AI is changing the education industry [13].

There are several AI applications for education. Popular areas to incorporate AI technologies to facilitate students' learning include engineering education, higher education, mathematics education, language education, surgical education, robotics education, computer science education, STEM education, medical education, musical education, and science education [14]. Companies using AI in education to enhance the classroom include Nuance, Knewton, Cognii, Querium, Century Tech, KidSense, Carnegie Learning, Kidaptive, Blippar, Thinkster Math, Volley, and Quizlet. These companies are merging the organic and the artificial by applying AI tools to innovate how people are educated [15]. Figure 6.4 compares traditional education with AI-driven education [16].

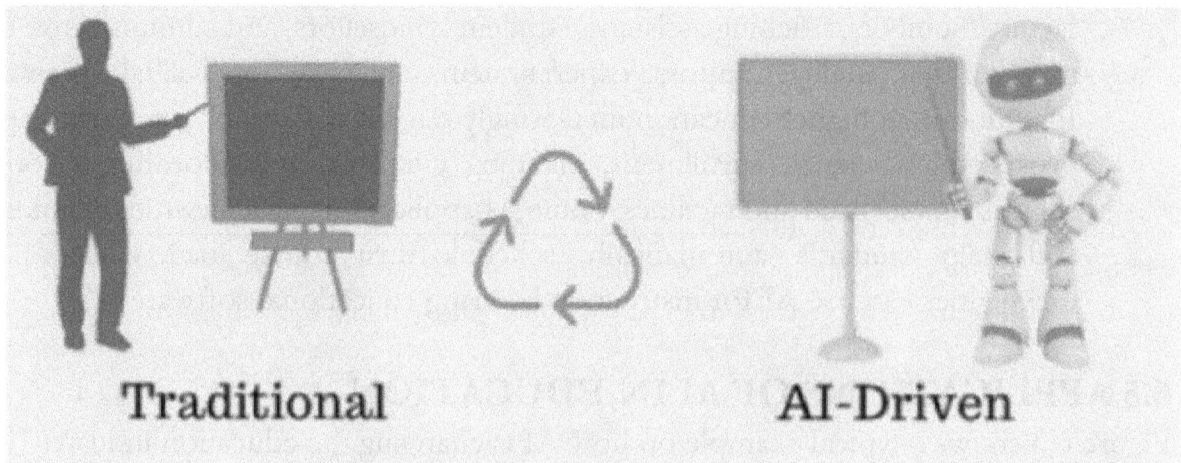

Figure 6.4 Comparing traditional education with AI-driven education [16]

An application is the use of expert systems to assist with educational diagnosis and assessment. Other applications include the following.

- **In the Classroom:** While AI can never replace human teachers, it can play a great role in the classroom. AI can allow teachers to hand off all assignments to an AI for grading so the teachers can spend more time with students. Despite the cost and need for Internet, AI is useful for tutoring. Since teachers cannot be available for students all the time, tutors are needed. Working with an AI tutor can help students with social or academic anxiety get the help [17].

- **Personalized Education:** Traditional education is not flexible. AI will enable personalized or tailored education. AI can provide a level of differentiation that customizes learning specifically to an individual student. Artificial intelligence helps build a personalized study schedule for each learner, thereby tailoring studies according to student's specific needs. This opens up new ways of interacting for students with learning disabilities. Personalized education increases efficiency, improves accessibility, and scales the processes [18].

- **Administration:** AI can simplify administrative tasks. It can automate the expedition of administrative duties for teachers and academic institutions. Technology can be used to automate the grading tasks where multiple tests are involved. This means that professors would have more time with their students rather than spending long hours grading them.

- **Universal Access to Global Classrooms:** AI can help to eliminate boundaries, thereby facilitating the learning of any course from anywhere, anytime across the globe. AI tools can help make global classrooms available to all including those who speak different languages. There will be a wider range of courses available online and with the help of AI, students will be learning from wherever they are [19].

- **Medical Education:** Essentially, medicine serves common human needs and promotes patient well-being. AI is critical for the future development of healthcare. The rate at which emerging health AI technologies are developing, being introduced into clinical practice, and being used by patients requires that healthcare professionals are well equipped. Physicians must work with patients to make the implementation of AI technologies transparent and accountable [20]. AI can be used to check the effectiveness of the medical program and satisfaction of the medical students. Today, AI is being applied in healthcare for faster and accurate diagnosis, to augment radiology, reduce errors due to human fatigue, reduce medical costs, replace repetitive, labor-intensive tasks, and reduce mortality rates [21].

- **Marketing Education:** Artificial intelligence is transforming marketing profession. It has several applications in marketing. These include sales forecasting, website experience personalization, speech recognition, content creation, chatbots, etc. For example, chatbot is regarded as one of the most promising AI applications. The virtual assistant is used in online marketing. Learning to comprehend and use AI can be beneficial to students. Marketing students believe that they will work with AI tools during their careers [22].

Other applications include personalized guidance, supports, feedback, assessment tools, remediation, virtual assistants for students, mobile games, intelligent tutoring systems, educational robots, smart education, engineering education, precision education, and design education.

6.6 GLOBAL ADOPTION OF AI IN EDUCATION

The wave of interest in artificial intelligence is already impacting the world economy and has captured the attention of many analysts. Global adoption of AI technology in education is transforming the way we teach and learn. Artificial intelligence helps assess what a student does and does not know. AI applications can perform assessment and evaluation tasks at very high accuracy and efficiency levels. Tech giants like Apple, Google, Microsoft, and Facebook currently compete in AI industry and are investing heavily in new applications and research.

- **United States:** AI in education generally focuses on identifying what a student does and does not know through diagnostic testing and then developing personalized curricula based on each student's specific needs. AI-based education platforms in US include Carnegie Learning, ALEKS, Jill Watson, Alexas, Siri assistant, and Google Homes. Solutions for human-AI interaction and collaboration are already available to help people with disabilities. At the beginning of 2017, Einstein robots were born in the United States.

- **China:** The Chinese have been working on creating smart education. In China, systems are already being used to monitor student participation and expressions via face recognition in classrooms. The Chinese government has set 2030 as the deadline to integrate AI with the Chinese infrastructure. The government's ambitious plan would require huge amounts of research in AI, supported by professionals trained in the technology. China is investing billions in AI and a significant portion center around education technology. It has started a grand experiment in AI education, which could reshape how the world learns [23]. Toward the end of 2016, an elementary school introduced the cute robot "Xiao'pang" into the classroom and interacted with children.

- **Sweden:** In Sweden, Sana Labs takes an entirely different approach to building AI for education. Rather than creating their own platform, Sana actually helps other companies personalize education by taking care of the AI and machine learning side of things.

- **Canada:** Artificial intelligence has made few inroads into Canadian education systems. AI is already being used for learning in Québec and the rest of Canada. Québec launched its Digital Action Plan for Education and Higher Education in May 2018. It states that, "The shift to digital is a unique opportunity for the development and growth of Québec and that digital technologies play a role in the educational success of our young people by offering them new ways to learn, communicate, share, create and collaborate in short, by breathing new life into our schools." Some experts worry about the risks of AI for society if the forward march of intelligent machines is left unchecked [24].

- **United Kingdom**: A multitude of AI-driven applications are already in use in schools and universities. Many recent ITS use machine learning techniques, self-training algorithms, and neural networks, to enable them to make appropriate decisions about what learning content to provide to the learner. Despite nearly three decades of work, the benefits and enormous potential of AI remain mostly unrealized. Some within the scientific community worry that AI is a Pandora's box with dangerous consequences [25].

6.7 BENEFITS

AI tools are gradually changing the landscape of education. They can help make global classrooms available to all. AI-based tools have a high potential to support students, faculty members, and administrators throughout the student lifecycle. Today, AI applications in education (AIED) are widely used by learners and educators. Although AI provides many benefits for students and teachers, it cannot replace teachers. The benefits of AI for students, teachers, and school include [26,27]:

- **Education At Any Time:** AI is impacting on how children can learn and from where. AI-based applications allow students to study whenever they have free time and receive feedback from teachers in a real-time mode.
- **Virtual Mentors:** AI-based platforms offer virtual mentors to track the students' progress. AI holds promise as a tool to monitor student performance.
- **Better Engagement**: Modern technologies like VR and gamification help involve students in the education process, making it more interactive and personalized.

- **Equality**: The use of AI is also being framed as a potential boom to equality. All students will have equal access to educational resources and have equal opportunities.

- **Opportunity to Find A Good Teacher:** Educational platforms have a lot of teachers, so the student has an opportunity to communicate with specialists from other countries.

- **Personalized learning:** AI-based solutions offering personalized learning are about to transform the school curriculum as well as the entire education sector.

- **Teacher's Aid:** AI can be a great ally to a teacher. AI can help an educator reduce the burden of administrative duties such as marking exams, grading students' assignments, planning, etc. and save a lot of time.

- **Teaching the Teacher:** Artificial intelligence makes comprehensive information available to teachers any time of day.

- **Connecting Everyone:** AI tools can help make global classrooms available to all, fostering greater cooperation, communication, and collaboration among schools and nations.

- **Cost Reduction**: AI can accelerate the pace and reduce the cost of learning.

- **Improves Efficiency:** AI reduces the burden of repetitive tasks teachers and schools must deal with daily. For example, students can use the AI-powered tools to learn word pronunciations, meaning, and proper usage. From classroom interactions, coursework learning, and administrative processes, AI makes things better and efficient.

- **Competitiveness:** Some educators regard AI as instrumental to their institution's competitiveness. AI has the power to become an equalizer in education and a key differentiator for institutions that embrace it.

- **Inclusivity**: Designers of AI applications and platforms must design with inclusivity in mind. They must address various capabilities and motivations of individuals. They must also avoid biases and assumptions with end users in mind during the design process. Designing for inclusion must identify who might be excluded.

- **Automation**: The area with the biggest automation potential is teacher's preparation and administrative responsibilities. Automation focuses on making mundane tasks easy for both students and teachers. For example, automated grading will allow teachers the time and freedom to do other things that are more important. AI can automate grading for nearly all forms of multiple-choice testing. Some of the benefits of artificial intelligence in education (AIED) are illustrated in Figure 6.5.

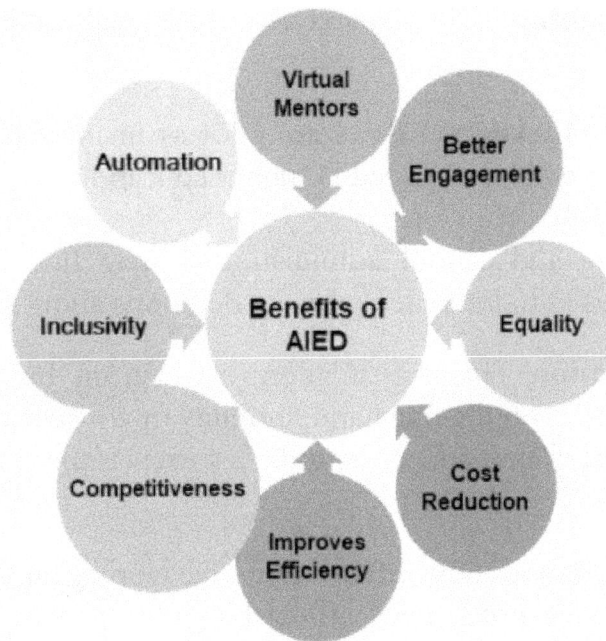

Figure 6.5   Benefit of artificial intelligence in education (AIED)

In AI-powered education, the world of academia will still need schools, classrooms, and teachers to motivate students and to teach social skills.

6.8 CHALLENGES

Although the possibilities of AI are exciting, a number of challenges prevent the full realization of AI in various educational institutions. Those challenges include [28]:

- **Privacy**: Personal information is private. Many students do not want others to know their private information, such as learning style and learning capability.

- **Limited Capability**: The capabilities of artificial intelligence in education are limited.

- **Unanswered Questions**: There are a lot of unanswered questions about AI's role and how it will be managed in higher education.

- **Trust**: Parents and school administrators may find it difficult to trust AI technologies used to influence or make decisions about student learning.

- **Ethical Dilemma**: This is a situation to which the human being is confronted when he opposes, on the one hand, morality or conscience and on the other, life. Another ethical dilemma faced by today's society is the "humanism" of the human being.

- **Cost**: The cost and time involved in developing and introducing AI-based methods may not be affordable by many public educational institutions. However, prices will come down with technology spread.

- **No Personal Connection:** Machines lack social skills and personal connection. They cannot teach empathy, compassion, and other such emotions, which are an integral part of the overall development of personality. Love and emotion are among the human characteristics that cannot be replaced by machines. A degree of objectivity inherent in AI technologies cannot handle subjective data accurately in the assessment tools common to educational aptitude tests.

When one compares the pros and cons of artificial intelligence in education, there are clearly more benefits.

6.9 FUTURE OF EDUCATION

All human actions are based on anticipated futures. Although we cannot predict the future, we can use our current knowledge to imagine futures. The better we understand the present and the history that has created it, to understand the possibilities of the future. Many predictions about the future of AI have been based on extrapolations of historical technical development [29].

Two decades ago, no one had heard about Google, Facebook, YouTube, or Wikipedia. Today they are the most used digital tools on the planet. These companies have invested millions of dollars developing artificial intelligence in education (AIED) products joining well-established multimillion dollar funded AIED companies such as Knewton26 and Carnegie Learning.

Artificial intelligence is one of the emerging technologies that can pervade and alter every aspect of our life. The introduction and adoption of AI technologies in learning and teaching has rapidly evolved in recent years. Although AI is just emerging and young, there is little doubt that the technology is inexorably linked to the future of education. Advances in artificial intelligence open to new possibilities and challenges for teaching, learning, education organization, and governance. While some AI solutions remain dependent on programming, some have an inbuilt capacity to learn patterns and make predictions. AI solutions have the potential to structurally change educational administrative services, the realm of teaching and learning. AI has the potential to replace a large number of administrative staff and teaching assistants in higher education [30]. The impressive list of applications of AIED has made some specialists to claim that teachers will not be necessary in future.

AI will impact the education field in the areas of administration, instruction, and personalized and individualized learning applications. We look towards a future when AI tools will support teachers in meeting the needs of their students. Now is the best time for educational institutions to rethink their function and future relation with AI solutions. Combining the best of human and machine for the benefit of the students is the ultimate goal of artificial intelligence in education.

6.10 CONCLUSION

AI is impacting education in various ways, and this is expected to continue. AI has produced educational tools which have attracted attention for their potential to improve education quality and enhance teaching and learning methods. As AI educational solutions matures, AI will be able to help fill needs gaps. There is no denying the place of AI in modern teaching and learning. Although the critical presence of teachers is irreplaceable, there will be many changes to a teacher's responsibilities.

Education might be a slower in adopting AI, but the changes will continue. In the coming decades, AI will transform education. The future of education is intrinsically linked with developments on new technologies and computing capacities of the new intelligent machines. It is therefore expedient for educators and policymakers to explore the intersection of education and artificial intelligence.

Since students of today will work in an environment where AI is the reality, it is expedient that our academic institutions expose students to AI technology. AI skills must also be balanced with person-centered aspects of education to develop a more rounded leaders of tomorrow. More information on artificial intelligence in education can be found in the books in [31-43] and following related journals:

- Artificial Intelligence Review

- International Journal of Artificial Intelligence in Education

- International Journal of Applications of Fuzzy Sets and Artificial Intelligence

- Computers & Education: Artificial Intelligence

- International Journal of Learning Analytics and Artificial Intelligence for Education

REFERENCES

[1] **P. Dadhich**, "Impact of artificial intelligence on the current education system," September 2020,

https://wire19.com/impact-of-artificial-intelligence-on-education-system/#:~:text=The%20implementation%20of%20educational%20AI,more%20knowledge%20in%20multiple%20subjects.&text=The%20implementation%20of%20AI%20can,well%20as%20streamline%20administrative%20tasks.

[2] "Real world artificial intelligence applications in various sectors," https://techvidvan.com/tutorials/artificial-intelligence-applications/

[3] "Ten facts about artificial intelligence in teaching and learning," https://teachonline.ca/sites/default/files/tools-trends/downloads/ten_facts_about_artificial_intelligence_0.pdf

[4] **X. Chen et al.**, "Application and theory gaps during the rise of artificial intelligence in education," Computers and Education: Artificial Intelligence, vol.1, 2020.

[5] **M. N. O. Sadiku, T. J. Ashaolu**, and **S. M. Musa**," Artificial intelligence in medicine: A primer," International Journal of Trend in Research and Development, vol. 6, no. 1, January. -Feb. 2019, pp. 270-272.

[6] **Y. Mintz** and **R. Brodie**, "Introduction to artificial intelligence in medicine," Minimally Invasive Therapy & Allied Technologies, vol. 28, no. 2, 2019, pp. 73-81.

[7] **O. Oana, T. Cosmin**, and **N. C. Valentin**, "Artificial intelligence - A new field of computer science which any business should consider," "Ovidius" University Annals, Economic Sciences Series, vol. XVII, no. 1, 2017, pp. 356-360.

[8] "How artificial intelligence could benefit the future of education," August 2019, Unknown Source.

[9] **M. N. O. Sadiku**, A. E. Sh**adare**, and **S.M. Musa**," Digital natives," International Journal of Advanced Research in Computer Science and Software Engineering, vol. 7, no. 7, July 2017, pp. 125-126.

[10] "The role of artificial intelligence in the future of education," March 2019, https://www.getsmarter.com/blog/market-trends/the-role-of-artificial-intelligence-in-the-future-of-education/

[11] **H. Burgsteiner**, **E. Allee**, and **M. Kandlhofer**, "IRobot: Teaching the basics of artificial intelligence in high schools," Proceedings of the Sixth Symposium on Educational Advances in Artificial Intelligence, 2016, pp. 4126-4127.

[12] **L. Plitnichenko**, "5 Main roles of artificial intelligence in education," May 2020, https://elearningindustry.com/ai-is-changing-the-education-industry-5-ways

[13] https://elearningindustry.com/ai-is-changing-the-education-industry-5-ways

[14] **X. Chen**, **H. Xie**, and G. **Hwang**, "A multi-perspective study on artificial intelligence in education: grants, conferences, journals, software tools, institutions, and researchers," Computers and Education: Artificial Intelligence, vol.1, 2020.

[15] **A. Schroer**, "12 Companies using AI in education to enhance the classroom," March 2020, https://builtin.com/artificial-intelligence/ai-in-education

[16] https://www.educationworld.in/how-ai-in-education-can-dominate-in-2020/

[17] **A. Sears**, "The role of artificial intelligence in the classroom," April 2018, https://elearningindustry.com/artificial-intelligence-in-the-classroom-role

[18] **S. Maghsudi et al.**, "Personalized Education in the AI Era: What to Expect Next?" https://arxiv.org/pdf/2101.10074.pdf

[19] **B. Marr**, "How is AI used in education -- Real world examples of today and a peek into the future," July 2018, https://www.forbes.com/sites/bernardmarr/2018/07/25/how-is-ai-used-in-education-real-world-examples-of-today-and-a-peek-into-the-future/?sh=2aea5075586e

[20] **V. Rampton, M. Mittelman**, and **J. Goldhahn**, "Implications of artificial intelligence for medical education,"
https://www.thelancet.com/journals/landig/article/PIIS2589-7500(20)30023-6/fulltext

[21] **K. Paranjape et al.**, "Introducing artificial intelligence training in medical education," JMIR Medical Education, vol 5, no 2 (2019): July-December. 2019.

[22] **S. Elhajjar, M. S. Karam**, and **S. Borna**, "Artificial intelligence in marketing education programs," Marketing Education Review, 2020.

[23] **A. Gupta**, "The role of artificial intelligence in education," June 2020, https://discover.bot/bot-talk/role-of-artificial-intelligence-in-education/

[24] **T. Karsenti**, "Artificial intelligence in education: The urgent need to prepare teachers for tomorrow's schools," Formation et Profession, vol. 27, no. 1, 2019, pp. 105-111.

[25] **R. Luckin and W. Holmes**, "Intelligence unleashed: An argument for AI in education,"

https://static.googleusercontent.com/media/edu.google.com/en//pdfs/Intelligence-Unleashed-Publication.pdf

[26] **V. Kuprenko**, "Artificial intelligence in education: Benefits, challenges, and use cases,"

https://medium.com/towards-artificial-intelligence/artificial-intelligence-in-education-benefits-challenges-and-use-cases-db52d8921f7a

[27] **"The role of artificial intelligence in the future of education,"** March 2019, https://www.getsmarter.com/blog/market-trends/the-role-of-artificial-intelligence-in-the-future-of-education/

[28] **O. Zawacki-Richter et al.**, "Systematic review of research on artificial intelligence applications in higher education – Where are the educators?" International Journal of Educational Technology in Higher Education, 16, vol. 16, no. 39, 2019.

[29] **I. Tuomi**, The Impact of Artificial Intelligence on Learning, Teaching, and Education. Policies for the future, EUR 29442 EN, Publications Office of the European Union, Luxembourg, 2018,

[30] **S. A. D. Popenici** and **S. Kerr**, "Exploring the impact of artificial intelligence on teaching and learning in higher education," Research and Practice in Technology Enhanced Learning, vol. 12, November 2017.

[31] Artificial Intelligence in Education; Building Technology Rich Learning Contexts That Work, (Int'l Conference on Artificial Intelligence in Education, 2007, Los Angeles, CA). IOS Press, 2007.

[32] **R. Luckin, K. R. Koedinger**, and **J. Greer (eds.)**, Artificial Intelligence in Education: Building Technology Rich Learning Contexts that Work IOS Press, 2007.

[33] **C. Fadel, W. Holmes**, and **M. Bialik**, Artificial Intelligence in Education: Promises and Implications for Teaching and Learning. Independently Published, 2019.

[34] **C. K. Looi et al.**, Artificial Intelligence in Education (Frontiers in Artificial Intelligence and Applications). IOS Press, 2005.

[35] **U. Hoppe, M. F. Verdejo**, and **J. Kay** (eds.), Artificial Intelligence in Education: Shaping The Future of Learning Through Intelligent Technologies. IOS Press, 2003.

[36] **J. E. Aoun**, Robot-Proof: Higher Education in the Age of Artificial Intelligence. The MIT Press, 2017.

[37] **R. Luckin**, Machine Learning and Human Intelligence: The Future of Education for the 21st Century. UCL Institute of Education Press, 2018.

[38] **R. W. Lawler** and **M. Yazdani**, Artificial Intelligence And Education: Learning Environments and Tutoring Systems. Intellect Ltd, 1987.

[39] **C. Fadel**, **W. Holmes**, and **M. Bialik**, Artificial Intelligence in Education: Promises and Implications for Teaching and Learning. Independently published, 2019.

[40] **G. Gauthier** and **C. Frasson (eds.),** Intelligent Tutoring Systems: At the Crossroad of Artificial Intelligence and Education. Intellect Ltd, 1990.

[41] **R. M. Cameron,** A.I. - 101: A Primer on Using Artificial Intelligence in Education. Exceedly Press, 2019.

[42] **U. Kose** and **D. Koc (eds.)**, Artificial Intelligence Applications in Distance Education. Information Science Reference, 2015.

[43] **R. J. Spiro**, **B. C. Bruce**, and **W. F. Brewer**, Theoretical Issues in Reading Comprehension: Perspectives from Cognitive Psychology, Linguistics, Artificial Intelligence and Education. Taylor & Francis, 2017.

CHAPTER 7
AI IN BUSINESS

"We're seeing a kind of a Wild West situation with AI and regulation right now. The scale at which businesses are adopting AI technologies isn't matched by clear guidelines to regulate algorithms and help researchers avoid the pitfalls of bias in datasets. We need to advocate for a better system of checks and balances to test AI for bias and fairness, and to help businesses determine whether certain use cases are even appropriate for this technology at the moment." — *Timnit Gebru*

7.1 INTRODUCTION

Artificial intelligence (AI) is becoming commonplace in our daily lives. It is now a household name. Amazon's virtual assistant Alexa may soon be in every home in America. AI is already disrupting virtually every business process in every industry.
AI has become an important technology that supports daily social life and economic activities. In recent years, AI has attracted attention as a key for economic growth in developed countries such as the United States and United Kingdom and developing countries such as China and India [1].

Artificial intelligence (AI) may be regarded as the intelligence that machines exhibit by imitating human behavior. It focuses on machine learning and enabling software to solve problems in a manner similar to human intelligence. For example, computers approve American Express purchases, diagnose faults for GE, and assign rooms for Holiday Inns. Voice recognition, computer vision, and robotics all employ AI [2].

Businesses of all types and sizes are considering artificial intelligence to solve their problems. The scope of AI in business transformation is constantly growing. AI technologies are drastically influencing the retail industry and customer experience. Some types of artificial intelligence are predominant in business, while others are not. AI systems are designed to make decisions using real-time data. They have the ability to learn and adapt as they make decisions. AI can increase productivity, gain competitive advantage, compliment human intelligence. and reduce cost of operations. Some of the top companies in the world have radically adopted the use of AI include Alibaba, Uber, Amazon, Tesla, and Microsoft [3].

This chapter provides an introduction to the applications of AI in business. It begins by briefly reviewing AI and what it takes for a business to adopt AI-based solution. It covers some applications of AI in business. It highlights some benefits and challenges of AI in the business realm. It discusses the future of business as far as AI is concerned. The last chapter concludes with comments.

7.2 REVIEW ON ARTIFICIAL INTELLIGENCE

Artificial intelligence (AI) is a field of computer science that is dedicated to developing software dealing with intelligent decisions, reasoning, and problem solving. Artificial intelligence is already part of our lives, slowly shaping our society and business. It is everywhere, in on your smartphones, laptops, and cars.

Artificial Intelligence may be regarded as any software program that depicts human intelligence and consequently has the ability to engage in a humanlike activity. AI tools can enhance decision-making abilities, allowing enterprises to perform increasingly complex tasks. They are versatile tools that enable people to rethink how we integrate information, analyze data, and use the resulting insights to improve decision making.

AI is not a single technology but a range of computational models and algorithms. AI is a collection of programmed algorithms to mimic human decision making. The major disciplines in AI include:

- Expert systems
- Fuzzy logic
- Neutal networks
- Machine learning

- Deep learning
- Natural Language Processors
- Robot

These branches of AI are illustrated in Figure 7.1.

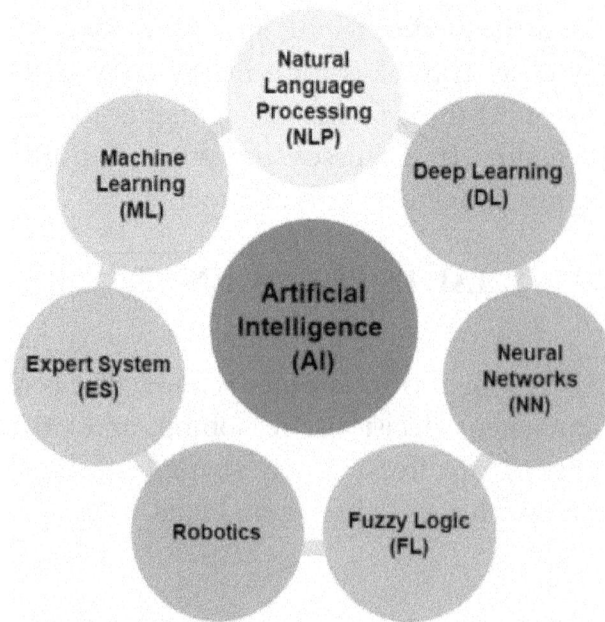

Figure 7.1 Branches of artificial intelligence.

All AI systems rely on algorithms, which are basically a set of instructions on how to organize and manage data. These computer-based tools or technologies have been used to achieve AI's goals. Each AI tool has its own advantages. Using a combination of these models, rather than a single model, is recommended.

7.3 BEFORE ADOPTING AI

Artificial intelligence is already widely used in business applications, including automation, data analytics, and natural language processing. Regardless of the kind of business you operate, you are in it to make profits. To increasing your profitability, artificial intelligence (AI) has proven itself to be the next big thing in business development. AI has the potential to revolutionize how you run your business. If you want to scale your business to grow beyond the current level, you need to think about how you can implement artificial intelligence in your business.

When a company decides to begin its AI journey, it often makes sense for analytics executives, data scientists, data engineers, user interface designers, visualization specialists to sit down and work together. Therefore, before adopting an AI strategy, companies should ask themselves the following five questions [4]:

1. What is the problem you plan to solve with AI?

Being clear about where the AI opportunities are and having defined strategies to obtain the data that AI requires should be the starting point for any business that want to implement AI. You should not expect AI to solve problems that you have not clearly identified. Your best bet is to solve current problems with technology already available. You should start with your business priority.

2. What is the company's plan to turn AI into an opportunity?

It is essential to know how to reformulate the problem definition in an automatic learning problem and how to implement it in a way that avoids any loss of value during the transformation process. How will AI be integrated with the company's overall strategy? Companies desiring to adopt AI should look at what already exists. New applications are being developed daily to leverage artificial intelligence in business. Organizations will differentiate by aligning strategic priorities with emerging technologies to drive innovation. AI adoption must be done in collaboration with IT and the entire business in order to apply it in a way that provides real benefits.

3. Does the company have the necessary digitized data to feed the AI model?

The quality of the AI model is directly dependent on the quality and quantity of data available to the company. Does my company have enough data? Are the data sources that the AI will use are reliable? Do I have the data stored in digital systems? To be able to manage the data correctly, they must be digitized, centralized, organized, periodically backed-up, and integrated in different digital tools.

4. Does the company have the necessary resources for the implementation?

The company must be realistic about whether it really has the necessary resources at the level of human and financial capital to adopt AI. The accuracy of the AI model will depend on the budget, equipment, and time available to the company to develop it. The technologies that enable AI are advancing rapidly and becoming increasingly affordable. It is important to align a company's culture and structure to support AI adoption. Business leaders must provide a vision that rallies all workers around a common goal. Given the risks involved in the adoption of AI-based tools, their adoption is currently limited to startups and large companies.

5. What are the expected returns from applying this technology?

How long will it take for the company to recover the investment? How much will the company's costs be reduced once AI is implemented? Integrating AI tools in a company implies both recurring and non-recurring costs and therefore an important investment. A common mistake among business leaders is viewing AI as a plug-and-play technology with immediate returns. To ensure a smooth adoption of AI, companies need to educate everyone, from the top leaders down. Leaders must provide incentives for positive change.

7.4 APPLICATIONS OF AI IN BUSINESS

AI adoption must be done with IT and the entire business working together. It does not generally replace human intelligence but serves as a supporting tool. AI seems destined to profoundly impact all aspects of business. It is widely used in business applications such as marketing, finance, accounting, human resources, supply chain, automation, data analytics, and natural language processing. Some of these are explained as follows [5-9]:

- **Marketing:** Marketing is one of the most notable areas of applications of AI. It may decide the next most effective marketing strategy. AI helps to develop marketing strategies and execute them. One way AI is marketing products is through chatbots, which can help solve problems, suggest products/services, and support sales. Figure 7.2 shows some of the benefits using chatbots in business.

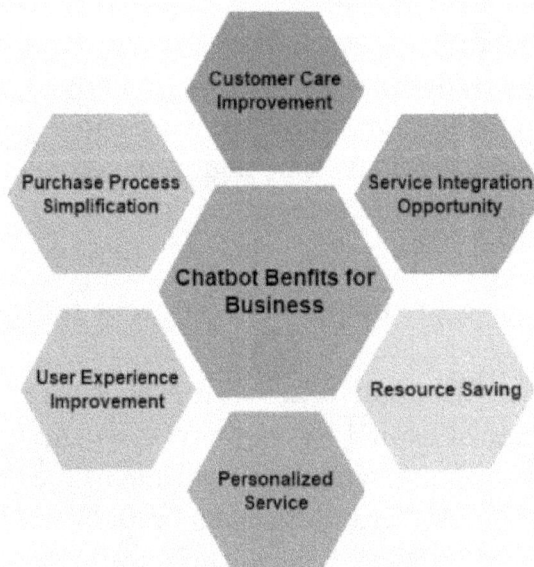

Figure 7.2 Some benefits of using chatbots in business.

These AI chatbots can understand natural language. Retailers will be able to use AI technology to track shoppers in their stores and prevent theft. For example, Walmart has been using HANA to process high volume of transaction records.

Finance: Investments in financial AI is growing as AI is increasingly being applied in financial institutions such as banks and insurance companies. Some banks use various AI tools to detect fraudulent activity. Loan applications are now being processed by software that can take into account a variety of factors such as creditworthiness, credit score, and background check. AI could make an educated guess and predict how likely a person is to pay back a loan based on their history and give the lender a recommendation. Fraud detection represents another way AI is helpful in financial systems since AI can discern fraudulent activities.

Accounting*:* AI is redefining the job descriptions of all sectors of professions. Accounting is a discipline that deals with recording, analyzing, summarizing, and reporting every business transaction. Accounting can be divided into several parts based on the activity of the organization: financial accounting, cost accounting, and management accounting. The AI technology is revolutionizing the accounting field and altering the roles of accountants [10].

Automation*:* The future is gravitating towards automation with AI driving the force behind eliminating or at least minimizing the human error from business operations. AI can help improve efficiency and save money by automating many tasks. Automation alleviates repetitive or dangerous tasks. The increasing penetration of AI and autonomous devices into many aspects of life is improving efficiency and response times. For example, robotic process automation can produce amazing results in accounting department. The Apptus eSales solution is designed to automate merchandising based on a predictive understanding of consumers. AI in accounting will reduce errors and free up professionals from repetitive tasks. Business communication with customers can also be automated through online chats, email marketing, and social media. Figure 7.3 illustrates robotic process automation [11].

Figure 7.3 Robotic process automation [11].

Human Resources*:* AI can help human resources (HR) departments by making candidate screening and recruitment process easier. Chatbots can be used to answer many commonly asked questions about company mission, policies, and benefits. Vendors in the HR sector such as Entelo, Textio, Textkernal, and HiringSolved offer AI solutions that help recruiters to sort and match potential workers and use bots to schedule candidate interviews. There is a notable impact from using AI in their staffing and talent management operations. AI can be a powerful tool in the hiring process.

- **Customer Service***:* The replacement of humans with AI in customer service is controversial. AI technology along with big data can be used to improve the customer experience. AI technology has automated certain customer service tasks. This technology is capable of adapting to the changing needs of the organization without increasing employee workload. Advances in language processing now allow AI to communicate directly with customers and employees alike. With helpdesk chatbots in place, customers can interact with companies in real-time to resolve complaints, place orders, get information, and do almost anything else.

- **Intelligent Supply Chain***:* As consumer expectations continue to change, supply chains struggle to get the right products to customers, when and where they need them, without impacting margins. AI enables a self-learning and self-optimizing supply chain model with real-time insights to deliver precision.

- **Business Intelligence**: This may also be regarded as a collection of decision support technologies for the enterprise aimed at enabling professionals make better and faster decisions. The purpose of business intelligence (BI) is to support better business decision-making [12]. The use of AI and machine learning in BI is helping business enterprises to elicit actionable insights from complex data. Machine learning tools in business intelligence like the HANA will improve operational efficiency. BI chatbots are enabling decision making by analyzing business data. AI has merged with BI applications in manufacturing and industrial sectors. Figure 7.4 shows a typical healthcare business intelligence [13].

Figure 7.4   Healthcare business intelligence [13].

- **Predictive Analytics**: Predictive analytics, as the name suggests, is the type of analysis that uses data, algorithms, and various techniques to forecast the future as a function of the past. Using predictive analysis, AI allows businesses to anticipate their users' needs and offer them solutions to those needs beforehand. AI can help businesses to identify who is most likely to buy their product. For example, banks use AI for predictive analytics, in fraud detection, and in determining customers that are likely to repay loans before approving them.

- **Cybersecurity**: As cyber-attacks increase and more sophisticated tools are used to breach cyber defenses, human operators are no longer capable. Artificial intelligence can analyze networks to identify vulnerabilities and breaches for businesses. This type of network security is superior to previous forms, such as firewalls. Artificial intelligence is an ally when it comes to looking for holes in computer network defenses. AI solution can monitor behavior, detect anomalies, and respond to threats. AI has become an indispensable tool in a business' cybersecurity infrastructure.

These are just some of the examples of AI applications in business. Other uses of AI in business include business analytics, chatbots and virtual assistance, fraud/crime detection, sales, R&D, ecommerce, personalized advertising, customer service, talent selection and development, manufacturing, healthcare business, national security, criminal justice, logistics, transportation, and smart cities.

7.5 BENEFITS

Most business leaders are excited about incorporating AI into the company's business functions in order to start realizing its extraordinary benefits. Companies of all types and sizes are finding ways to use the right AI technology to save time and money. In order to meet evolving customer expectations, companies are leveraging intelligent technologies to transform the way they operate.

The business benefits of artificial intelligence are many. By deploying the right AI technology, a business can enjoy the following benefits [14]:

- Save time and money by automating routine or repetitive processes

- Reduce operational costs, increase efficiency, boost operational efficiency, and improve customer experience

- Increase sales, detect fraud, improve human resources, and provide predictive analysis

- Increase productivity and operational efficiencies and drive revenue growth

- Predict customer preferences and offer them better, personalized experience

- Increase revenue by identifying and maximizing sales opportunities

- Provide competitor advantage and empowers businesses to identify new opportunities

- Meet customer's demands, impulses, and habits of purchasing

- Enable companies to do things differently, enable better decision making, create more intelligent processes, and generate insights

- Augment rather than replace human capabilities.

- Improve personalized services and predict customer needs with remarkable accuracy.

- Enable businesses to work smarter and faster, doing more with less.

- Help customer service reps provide better support.

- Personalize learning and improve accuracy of knowledge worker.

AI-based devices like chatbots and virtual assistants are on the rise. Amazon's Alexa, Google's Home, Apple's Siri, and Microsoft's Cortana are all using AI-based algorithms to make life better.

7.6 CHALLENGES

Every new technology comes with risks. As an emerging technology, AI is changing at a fast pace and may present some unexpected challenges. Some of the challenges facing business AI include the following [15,16].

- **Lack of familiarity**: Most people in business are not very familiar with AI, what it is and what it can do for them. In spite of this, AI is a technology that is transforming every walk of life.

- **Misunderstanding**: There is a lot of misunderstanding in the business world about AI's current capabilities and future potential. AI is misunderstood by many, especially by the mainstream media. The media has overpraised AI for techniques that are not new and over-criticized it for overly optimistic promises.

- **Public Fear**: Everywhere you look, it seems AI is assisting and displacing human effort. There are plenty of doom and gloom predictions around AI. AI offers both promise and peril as it revolutionizes the workplace. AI has created widespread fear of job losses and further rises in inequality. There is a public fear around the world particularly in the business community that AI technology such as robots will overtake us and force humans into obsolescence through nefarious cyber-attacks. This fear is considered unfounded by some. For sure, AI is nowhere near replacing humans, but it will automate repetitive tasks and free us up to do more complex tasks.

- **Ethics**: There are currently no standards concerning data access, data sharing, or data protection. How should we promote data access? How do we guard against biased or unfair use of data in algorithms? What types of ethical principles are introduced through software programming? To answer these and related questions it may require an international body that will set the standards by which ethical dilemmas are resolved. The IEEE Global Initiative has ethical guidelines for AI and autonomous systems.

- **Shortage of Workforce**: Due to the fast-growing AI market, people with AI skills are in short supply. To realize the full capacity of AI, we need the right people and the right culture working by the side of AI. Right now, there are shortages of data scientists, computer scientists, engineers, and software developers because students are not receiving instruction in AI skills. More emphasis should be put on STEM subjects (science, technology, engineering and mathematics). Not generating more people with these capabilities will limit AI development.

AI initiatives face formidable cultural and organizational barriers. Some obstacles, such as workers' fear of becoming obsolete, are common across organizations.

While AI offers valuable benefits to businesses, implementation is usually expensive and time-consuming. Overall, the pros outweigh the cons. It is hoped that the present culture would undergo a remarkable evolution till the time is ripe and then AI would permeate all areas of life in all its fullness,

7.7 FUTURE OF BUSINESS

As mentioned earlier, artificial intelligence has a wide range of uses in the business world. Businesses seem to be entering a new era ruled by data. AI will continue to shape the business world for years to come in ways we can hardly imagine. Whether rosy or rocky, the future is coming quickly, and artificial intelligence will certainly be a part of it. People are always interested in knowing what will happen next and this is an important part of business. Experts are already making predictions on what is coming. Here are four exciting trends to watch out for coming years:

Competitive Edge: AI is enabling businesses to work smarter and faster, doing more with significantly less. They have been eager to adopt AI-powered tools to simplify complicated processes, meet the needs of customers, and maintain a competitive edge. Businesses that do not use AI to their advantage will soon be left behind. Businesses that excel at implementing AI will find themselves at a great advantage in a world where humans and machines working together to outperform either humans or machines working on their own.

Automation: Although it is difficult to predict how the technology will develop, most experts see that robots will become extremely useful in daily life. AI will create more wealth than it destroys. Companies in developing countries may use AI-based solutions to increase their productivity, enhance autonomous goods and service delivery, implement production automation.

Business Intelligence: We predict that business intelligence applications will be one of the fastest growing areas for leveraging AI technology over the next five to ten years. In some industries, AI is capable of automating business intelligence and analytics processes, providing a holistic end-to-end solution.

Breakthroughs: As technology continues to advance, the ability to apply AI to various aspects of the organization will advance too. Certain things that we cannot do today will be possible in the future. AI has the potential to dramatically remake the economy. It will be a significant part of business strategy and enable smarter decision-making.

Businesses should be strategic in how they approach AI and adopt it into their existing business plan and workforce.

7.8 CONCLUSION

The concept of artificial intelligence is growing in the business world. AI has become a trend that cannot be disregarded. Today, artificial intelligence has become a household name. Basically, AI is the capability of a machine to imitate intelligent behavior. It has played a significant role in transforming the way business operate today. AI may well be a revolution in human affairs because AI tools will have substantial impact on the general public in the foreseeable future. Companies around the globe are utilizing AI in order to thrive, provide better solutions to their clients, and create new sources of business value.

We are in an era of AI. It is evident that AI is driving powerful transformations across a variety of industries. Business managers may not fully understand AI technology, but they need to understand what it can do for their business. They should learn to integrate AI technology into multiple aspects of their business. To get the most out of AI, data scientists must engage with business users to understand their needs and problems. More information on artificial intelligence in business can be found in the books in [17-37]. One may also consult a related journal: Artificial Intelligence Review.

REFERENCES

[1] **H. Lu et al.**, "Brain intelligence: Go beyond artificial intelligence," Mobile Networks and Applications, *vol* 23, 2018, pp. 368–375.

[2] **P. J. Byrne** and **S. P. Franklin**. "Can your business use artificial intelligence?" Business Perspectives, vol. 3, no. 3, Spring 1990.

[3] **D. Parikh**, "Introduction to machine learning-studying about linear and logistic regression," June 2018,

https://medium.com/coinmonks/introduction-to-machine-learning-studying-about-linear-and-logistic-regression-434fdaf2f709

[4] "10 questions you should ask yourself before implementing artificial intelligence in your company," July 2020,

https://nexusintegra.io/artificial-intelligence-in-business/

[5] "Business applications for artificial intelligence: An update for 2020," March 2019,

https://blog.dce.harvard.edu/professional-development/business-applications artificial-intelligence-what-know-2019

[6] **B. Marr**, "10 Business functions that are ready to use artificial intelligence," March 2020,

https://www.forbes.com/sites/bernardmarr/2020/03/30/10-business-functions-that-are-ready-to-use-artificial-intelligence/#70fe2c923068

[7] **D. M. West and J. R. Allen** ,"How artificial intelligence is transforming the world," April 2018
https://www.brookings.edu/research/how-artificial-intelligence-is-transforming-the-world/

[8] "Why is artificial intelligence in business analytics so critical for business growth?" September 2019, https://medium.com/gobeyond-ai/why-is-artificial-intelligence-in-business-analytics-so-critical-for-business-growth-2ed16cbe7846#:~:text=AI%2Dpowered%20BI%20tools%20have,day%2Dto%2Dday%20decisions.

[9] "What are the benefits to a business of using a chatbot," https://www.techware.co.in/chatbotsforbusiness.php

[10] **D. Kwafo**, "The impacts of artificial intelligence on management accounting students: A case study at Oulu Business School, University of Oulu," Master's *Thesis*, University of Oulu, April 2019.

[11] "5 Examples of artificial intelligence in business applications," June 2020, https://usmsystems.com/artificial-intelligence-in-business-applications/

[12] **M. N. O. Sadiku, M. Tembely**, and **S. M Musa**," Toward better understanding of business intelligence," Journal of Scientific and Engineering Research, vol. 3, no. 5, 2016, pp. 89-91.

[13] S. **Y. Lee**, "Architecture for business intelligence in the healthcare sector," IOP Conference Series: Materials Science and Engineering, 2018.

[14] "Artificial intelligence in business: Business benefits of artificial intelligence," https://www.nibusinessinfo.co.uk/content/business-benefits-artificial-intelligence

[15] **D. Matskevich,** "Preparing your business for the artificial intelligence revolution," July 2018,

https://www.forbes.com/sites/forbestechcouncil/2018/07/12/preparing-your-business-for-the-artificial-intelligence-revolution/#39d5037d7ac8

[16] **A. Lauterbach** and **A. Bonime-Blanc**, "Artificial intelligence: A strategic business and governance imperative,"

https://gecrisk.com/wp-content/uploads/2016/09/ALauterbach-ABonimeBlanc-Artificial-Intelligence-Governance-NACD-Sept-2016.pdf

[17] **M. Gilbert (ed.)**, Artificial Intelligence for Autonomous Networks. Boca Raton, FL: CRC Press, 2018.

[18] **R. Akerkar**, Artificial Intelligence for Business. Springer, 2018.

[19] D. Rose, Artificial Intelligence for Business. Pearson Education, 2020.

[20] **J. L. Anderson**, and **J. L. Coveyduc**, Artificial Intelligence for Business: A Roadmap for Getting Starred with AI. John Wiley & Sons, 2020.

[21] **T. B. Cross**, Knowledge Engineering: The Uses of Artificial Intelligence in Business. Upper Saddle River, NJ: Prentice Hall, 1988.

[22] **J. M. Munoz** and **A. Naqvi**, Business Strategy in the Artificial Intelligence Economy. Business Expert Press, 2018.

[23] **A. Lauterbach** and **A. Bonime-Blanc**, The Artificial Intelligence Imperative: A Practical Roadmap for Business. Praeger, 2018.

[24] **J. Medicine**, Artificial Intelligence and Machine Learning for Business: Approach for Beginners to AI and Machine Learning and Their Revolution of Modern Life, Health Care, Business and Marketing. Independently published, 2019.

[25] **M. Skilton** and **F. Hovsepian**, The 4th Industrial Revolution: Responding to the Impact of Artificial Intelligence on Business. Springer, 2017.

[26] **R. Akerkar**, Artificial Intelligence for Business. Springer, 2019.
[27] S. Finlay, Artificial Intelligence and Machine Learning for Business: A No-Nonsense Guide to Data-Driven Technologies. Relativistic, 3rd edition, 2018.

[28] **O. Tensor**, Artificial Intelligence Business Applications. Oliver Tensor, 2020.

[29] **B. Mather**, Artificial Intelligence Business Applications: Artificial Intelligence Marketing and Sales Applications. Unknown Publisher, 2018.

[30] **M. Skilton** and **F. Hovsepian**, The 4th Industrial Revolution: Responding to the Impact of Artificial Intelligence on Business. Springer, 2017.

[31] **D. Partridge** and **K. M. Hussain**, Artificial Intelligence and Business Management. Norwood, NJ: Ablex Publishing, 1992.

[32] **W. R. Reitman** , Artificial Intelligence Applications for Business: Proceedings of The NYU Symposium, May, 1983. Norwood, NJ: Ablex Publishing, 1984.

[33] **W. B. Rauch-Hindin**, Artificial Intelligence In Business, Science, and Industry. Vol. II: Applications.   Prentice-Hall, 1985.

[34] **P. H. Winston** and **K. A. Prendergast (eds.)**, The AI Business: Commercial Uses of Artificial Intelligence. Cambridge, MA: MIT Press,1984.

[35] **M. Yao, A. Zhou**, and **M. Jia**, Artificial Intelligence: A Handbook for Business Leaders. TOPBOTS, 2018.

[36] **D. Rose**, Chicago Lakeshore Press, 2018.

[37] **J. Sterne**, Artificial Intelligence for Marketing: Practical Applications. John Wiley & Sons, 2017.

CHAPTER 8
AI IN INDUSTRY

"Just as electricity is fundamental to the way we live, in the not-so-distant future, it is not hard to see how AI will become the new electricity – embedded and/or supporting just about every aspect of our life." – *Daniel Eckert*

8.1 INTRODUCTION

Industry is continuously evolving and becoming more and more digitalized. As a result, data is continuously generated, processed, and analyzed. It would be great if the machines could gather insights from these volumes of data by themselves and optimize their processes. It is good to know that that is already being achieved, step-by-step, using artificial intelligence (AI). From driverless cars to personal assistant and virtual doctors, AI is transforming the way we live, work, travel, and do business. AI offers tremendous potential for industry [1]

Artificial intelligence may be regarded as the part of digital technology that denotes the use of coded computer software routines with specific instructions to perform tasks for which a human brain is usually considered necessary. Industrial AI usually refers to the application of artificial intelligence to industry. There are many reasons for the recent popularity of industrial AI: more affordable sensors, automated process of data acquisition, more powerful computation capability of computers, faster connectivity infrastructure, and more accessible cloud services [2]. The AI industry is driven by strong economic and political interests. AI is one of the basic concepts of the Industry 4.0 philosophy and a key technology for factory digitization. Industry 4.0 is a convergence of two worlds: Information Technology (IT) and Operational Technology (OT). The term "Industry 4.0" emerged from the German term "Industrie 4.0," which was first used in 2011 in a project sponsored by the German government that was focused on promoting computerization of manufacturing [3].

This chapter covers various applications of artificial intelligence in the industry. It begins by briefly reviewing AI. It discusses industrial AI and its various applications. It highlights the benefits and challenges of applying AI in industry. It addresses how AI is applied in industries worldwide. It covers the future of industrial AI. The last section concludes with comments.

8.2 REVIEW ON ARTIFICIAL INTELLIGENCE

In simple terms, Artificial intelligence (AI) refers to computer systems that mimic human cognitive functions. The term "artificial intelligence" (AI) was first used at a Dartmouth College conference in 1956. The main goal of AI is to enable machines to perform complex tasks that typically require human intelligence [4]. AI is now one of the most important global issues of the 21st century.

AI is not a single technology but a range of computational models and algorithms. AI is a collection of techniques that enables computer systems to perform tasks that would otherwise require human intelligence [5,6].   The major disciplines in AI include:

- Expert systems
- Fuzzy logic
- Neural networks
- Machine learning

- Deep learning
- Natural Language Processors
- Robots

These AI tools are illustrated in Figure 8.1.

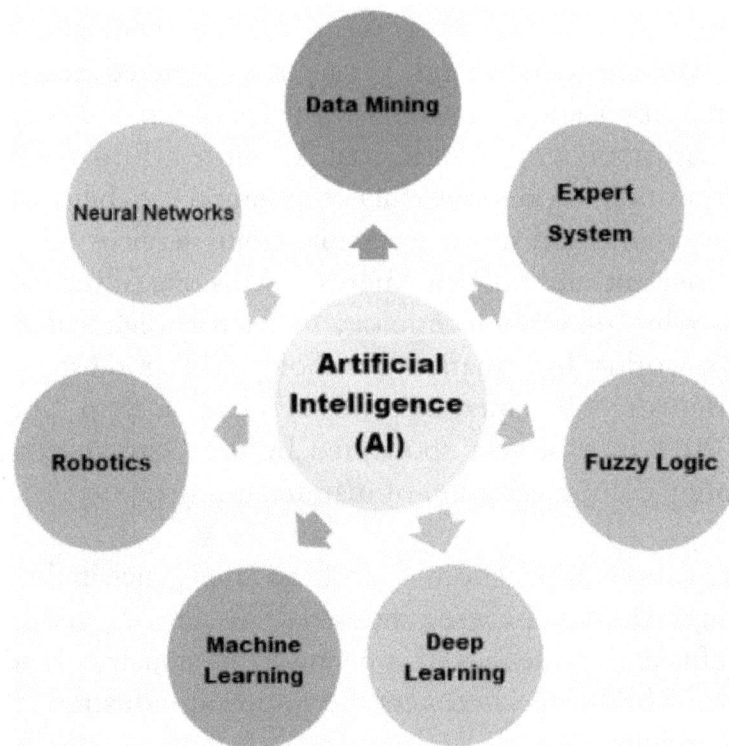

Figure 8.1 Branches of artificial intelligence.

Each AI tool has its own advantages. Using a combination of these models, rather than a single model, is recommended. AI systems are designed to make decisions using real-time data. They have the ability to learn and adapt as they make decisions.

AI-driven systems can discover trends, reveal inefficiencies, and predict future outcomes. These characteristics enable informed decision-making and AI to be potentially beneficial for many industries. Industrial AI deals with the application of AI technologies to address industrial issues such as customer value creation, productivity improvement, cost reduction, site optimization, predictive analysis, and insight discovery. An important feature of AI technology is that is can be added to existing technologies [7].

8.3 INDUSTRIAL AI

The current surge in AI research, investment, business, and industry applications is unprecedented. AI systems in industry are the same technologies you use in daily life but applied to industrial problems. Industrial AI (or IAI) can be embedded to existing products or services to make them more effective, reliable, safer, and to enhance their longevity. IAI tools can be classified into two categories: predefined *rules-based* tools and *machine learning* tools [8]. Rules-based IAI tools are ideal for well-understood processes that allow a small set of possible outcomes. Machine learning is often used in problems with vast amount of data because it needs examples and trials to determine correct behavior.

The key elements in Industrial AI can be characterized by "ABCDE." These key elements include Analytics technology (A), Big data technology (B), Cloud or Cyber technology (C), Domain knowhow (D) and Evidence (E). Analytics is the core of AI, which can only bring value if other elements are in place [9]. As shown in Figure 8.2, industrial AI has any application of AI relating to the physical operations or systems of an enterprise.

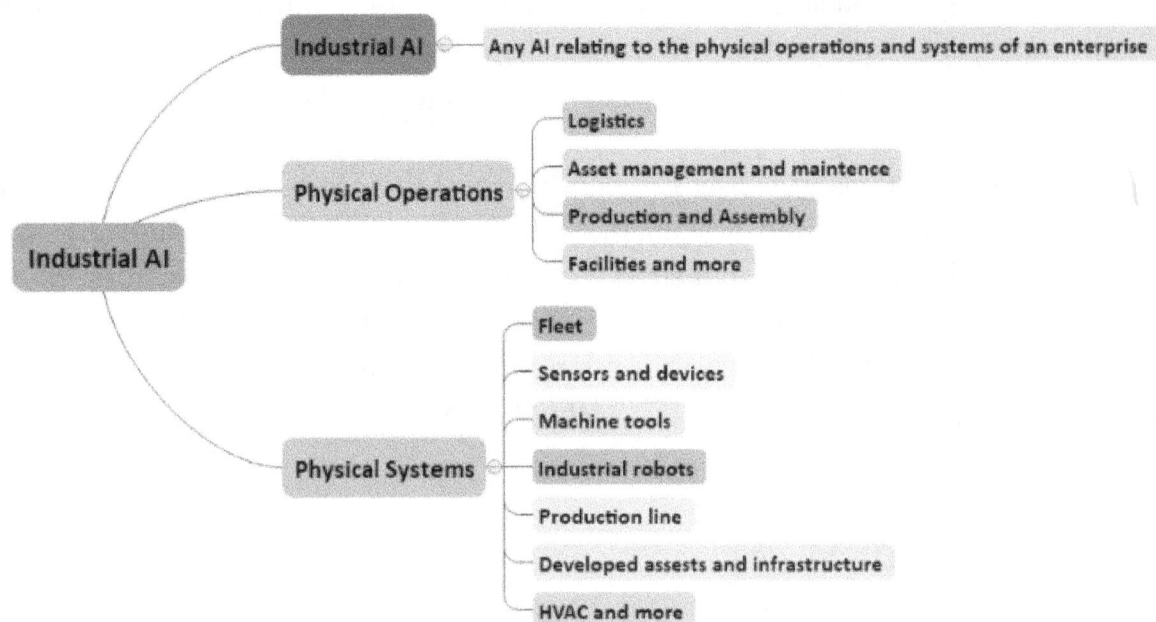

Figure 8.2   Industrial AI.

8.4 APPLICATIONS OF INDUSTRIAL AI

Artificial intelligence has impacted our lives significantly by fostering advances in many industries such as healthcare, ecommerce, pharmaceutical industry, energy industry, agriculture industry, petroleum industry, telecommunications industry construction industry, online advertising, consumer electronics, education, and entertainment. Each of the industries uses AI in revolutionary ways. The common applications of AI in major industries include the following [10-13]:

Manufacturing: This industry is undergoing unprecedented transformation driven by technologies that help manufacturers to digitize their factories. Without doubt, the manufacturing industry is leading the way in the application and adoption of AI technology. AI in the world of manufacturing has limitless potential. In manufacturing, AI is being employed across areas such as preventative maintenance, automation of human tasks, production, operations, and workforce planning. Robots serve as an integral part of the production process. AI algorithms are being used to notify manufacturing units of potential production faults that can lead to product quality issues. Figure 8.3 shows application of AI in manufacturing [14].

150

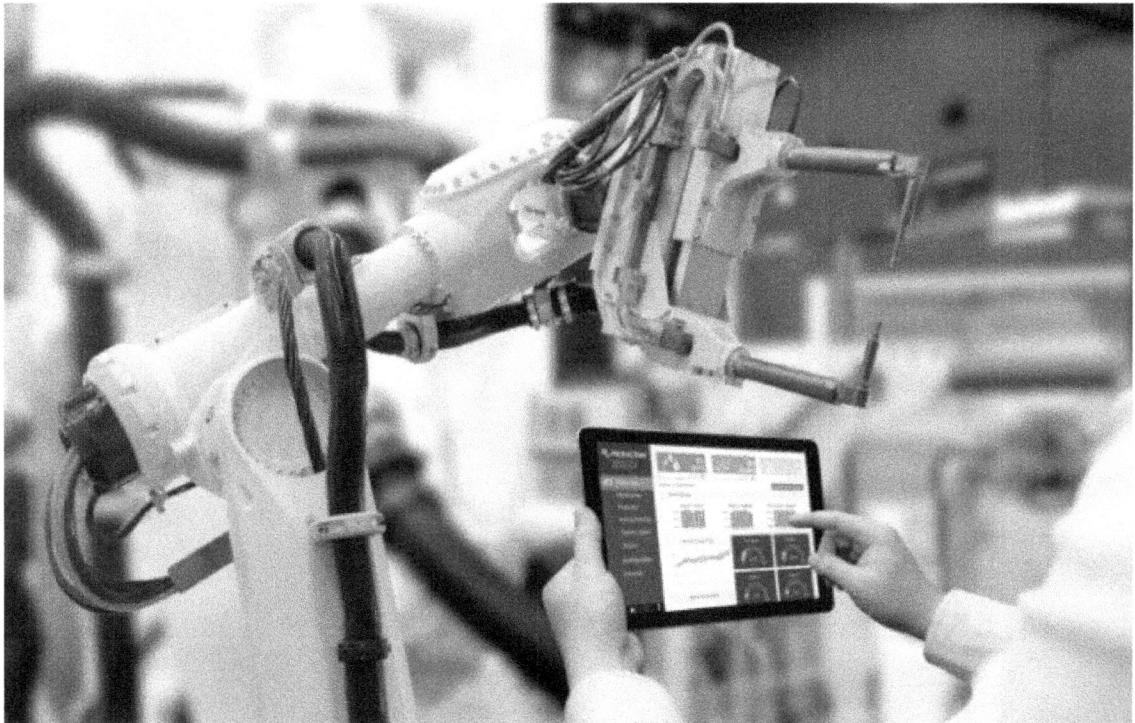

Figure 8.3 Application of AI in manufacturing [14].

- **Automotive Industry:** This industry is one of the largest benefactors of AI. Imagine no accidents, no traffic congestion, and no driver. Tesla and Google have already released self-driving cars, which were well received by their customers. People are already preferring pooling and shared rides over car ownerships. It is only a matter of time before self-driving cars go mainstream. AI-assisted self-driving cars make use of the sensors and cognitive equipment to drive safely, avoiding traffic and accidents. Three areas with the biggest AI potential are autonomous fleets for ride sharing, semi-autonomous features such as driver assist, and engine monitoring and predictive, autonomous maintenance.

- **Production:** AI technologies can improve the quality of industrial production, reduce costs, and simultaneously reducing production times. AI is making production more efficient, more flexible, and more reliable. It determines the production parameters that need to be adapted to ensure that this remains the case during the ongoing production process. As a result, production is made even more reliable and more efficient, thereby making companies even more competitive. In the petroleum and chemical industry, AI technology can effectively control the production process, optimize the process technology, improve production efficiency, and reduce energy consumption.

151

Automation: This is one of the major applications of industrial AI. A major goal of AI is to automate tasks that consume a lot of time if done by a human. From front desk receptionists at clinics to certified nursing assistants (CNAs), registered nurses (RNs), doctors, surgeons, and those in specialized areas, AI is helping automate processes in the healthcare industry. Robotics in the form of AI-powered bots assist with surgery. Using robots to assist surgeons helps with higher accuracy, lower risk of infection, less blood loss and pain, shorter hospital stay, and quicker recovery times. Figure 8.4 illustrates a robotic cardiac surgery [15].

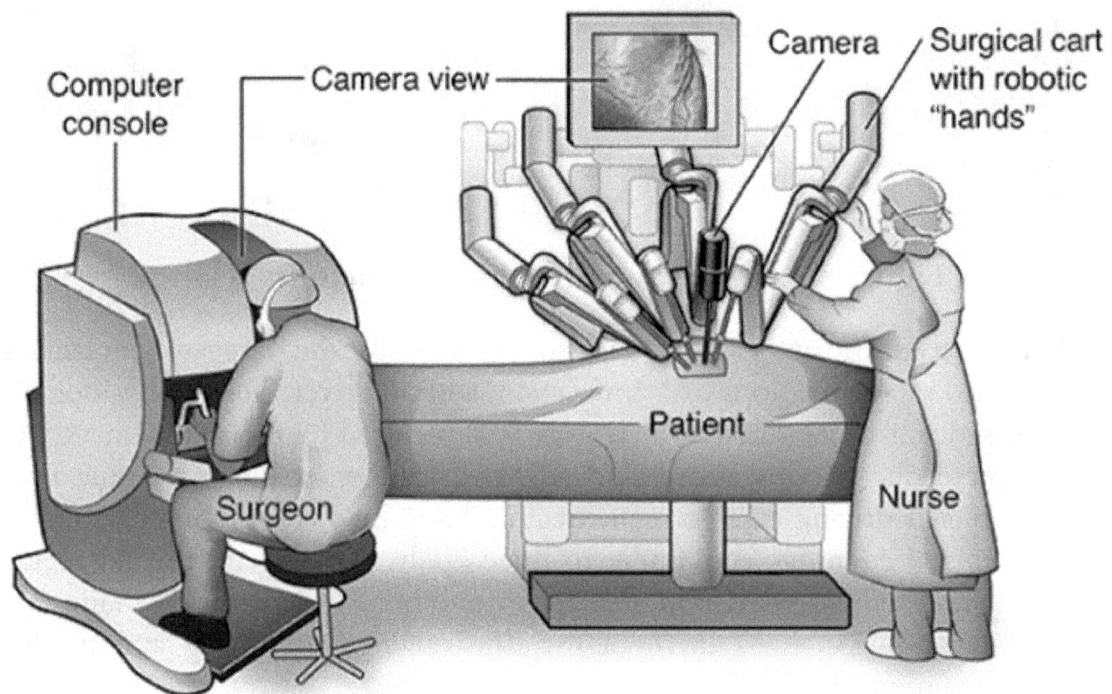

Figure 8.4   Robotic cardiac surgery [15].

Electronic health record (EHR) technicians are using AI to help automate digital medical records by documenting, sorting, and storing records in digital form. Ecommerce is another sector that benefits greatly from automation.

- **Healthcare:** AI is currently being applied for a wide range of healthcare services. The various ways AI-based programs help healthcare organizations include:
  - Improvements in medical imaging
  - More accuracy when reading test results
  - Robot-assisted surgeries
  - Administrative support
  - Efficient recruiting

Natural language processing (NLP) shows promises in drug safety. AI can be used to identify people at risk and recommend therapy before they fall into depression. The contribution of the technology giants like Microsoft, Google, Apple and IBM in the healthcare sector holds significant importance for the industry. In 2013, the MD Anderson Cancer Center initiated a project in 2013 for diagnosing and recommending treatment plans for certain forms of cancer using IBM's Watson cognitive system. Figure 8.5 displays application of AI in healthcare [16].

Figure 8.5 Application of AI in healthcare [16].

Retail and E-commerce: This industry will be one that is most impacted by AI. It may be the only area where the application of AI is the most observable to the majority of end-users because AI technologies are increasingly being used to enhance the customer experience and solve business problems. AI can support three types of business needs: automating business processes, gaining insight through data analysis, and engaging with customers and employees. AI-based tools benefit the e-commerce companies by way of automating data, stock, and inventory analysis that facilitate better forecasting of sales. Behavioral analytics coupled with AI surveillance can help identify, alert, and prevent theft and other malicious practices inside the store. AI in the marketing industry has been so successful that people now expect personalization in their marketing emails, advertising, and website experiences.

Food Industry: The food industry has always been at the forefront of adopting emerging technologies to improve the sector. Artificial intelligence, with the capacity to makes computers to learn from experience, is playing a predominant role in the food industry. AI has found several applications in the food industry such as farming/agriculture, food processing, food sorting, food packaging, etc. AI is poised to revolutionize the food industry.

Banking and Financial Services: This industry is undergoing a massive transformation due to AI applications. For example, human agents are being replaced by intelligent software robots for processing loan applications in fractions of a second. AI-based chatbots are being deployed in the insurance industry to improve the customer experience.

Travel: This industry is deriving significant benefits from the widespread use of AI-enabled chatbots. Chatbots are a proven means for improving customer service and engagement mainly because of their 24/7 presence and instant resolution of queries. Many large travel organizations are turning to AI companies and using machine learning and predictive analytics to build their own AI-based mobile apps and chatbots for improving the customer experience.

Real Estate: The application of AI in this industry is opening new opportunities for agents, brokers, and clients. Agents are becoming more efficient and effective, brokers are getting more strategic, and consumers are feeling empowered. AI-powered bots can operate 24/7 and help brokers, and agents find the perfect match for people looking to buy, rent or sell their properties.

Entertainment and Gaming: AI is helping program producers and broadcasters identify which shows or programs they should recommend to individual users. For example, AI helps Netflix and Amazon provide a more personalized experience to users. In the film industry, AI is being employed to enhance digital effects in movies. In the music industry, large companies like Apple and Spotify implement AI to understand users' engagement patterns and recommend the right music to the right people at the right time [17].

Telecommunications Industry: This industry has a natural monopoly character. But things have changed since the 1980s; globalization and technology have all damped monopoly and encouraged competition in the industry. Today, the telecommunications industry is one of the most important areas for AI applications. Different AI techniques have their unique applications in the telecommunications industry. Expert systems and machine learning are the two AI techniques that have been widely used in telecommunications. Expert systems were designed for diagnosing complex equipment in an off-line mode. Software systems in telecommunications have to cope with a great variety of telecommunication protocols, and numerous hardware platforms and network architectures [18].

Chemical Industry: This industry is a complex socioeconomic system that is very difficult to manage. The industry is managed according to three main indices: production quantity, production quality, and sales price. It is a fertile ground for applying and developing AI technology. Areas of applications of AI and expert systems include process control, chemical synthesis and analysis, manufacturing, planning and configuration, waste minimization, and signal processing. AI technology can effectively control the production process, optimize the process technology, improve production efficiency, and reduce energy consumption [19].

Cybersecurity: There are countless ways in which the adversary can attack. Cybersecurity is the process of protecting computer networks from cyber-attacks or unintended unauthorized access. It is the need of the hour. Organizations, businesses, and governments need cybersecurity solutions because cyber criminals pose a threat to everyone. Artificial intelligence promises to be a great solution for this. By combining the strength of artificial intelligence with cybersecurity, security experts are more capable to defend vulnerable networks and data from cyber attackers. AI will be a game-changer in how we improve our cyber-resilience. Various AI tools have been increasingly applied for cybercrime detection and prevention.

There are the industries that have the highest potential for AI applications. Some of them are illustrated in Figure 8.6.

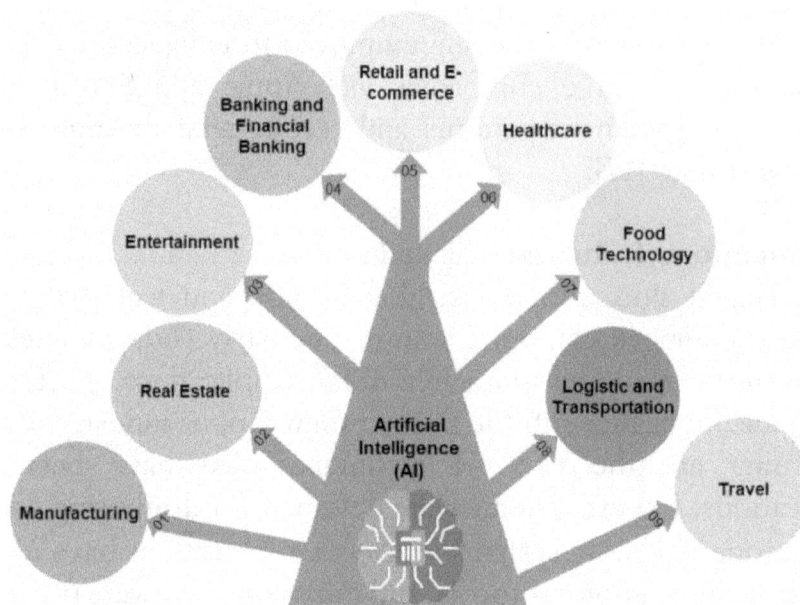

Figure 8.6 Some of the AI applications disrupting industries.

Other industries that can benefit from AI include pharmaceutical industry, energy industry, oil and gas industry, print industry, media industry, tour and travel industry, apparel/textile industry, aviation industry, hospitality industry, fashion industry, marketing, construction industry, education, legal, diagnostics, business intelligence, city planning, supply chain management, gaming industry, public relations industry, human resources, and the list continues.

8.5 BENEFITS

Industry holds a positive view of AI and sees AI transformation of economy unstoppable. AI is already making production more efficient, more flexible, and more reliable. Some benefits of AI include directed automation, 24/7 production, safer operational environments, flexible and efficient production, and reduced operating costs. Adoption of AI is moving fast paced across industries, where AI will clearly automate many processes that were earlier executed manually. Artificial intelligence empowers organizations to shift from reactive maintenance to predictive maintenance strategies.

Industrial AI has fundamentally changed the scope and pace of automation. AI applications in industry has improved the human experience as a whole [20,21]. Figure 8.7 shows the pros and cons of artificial intelligence.

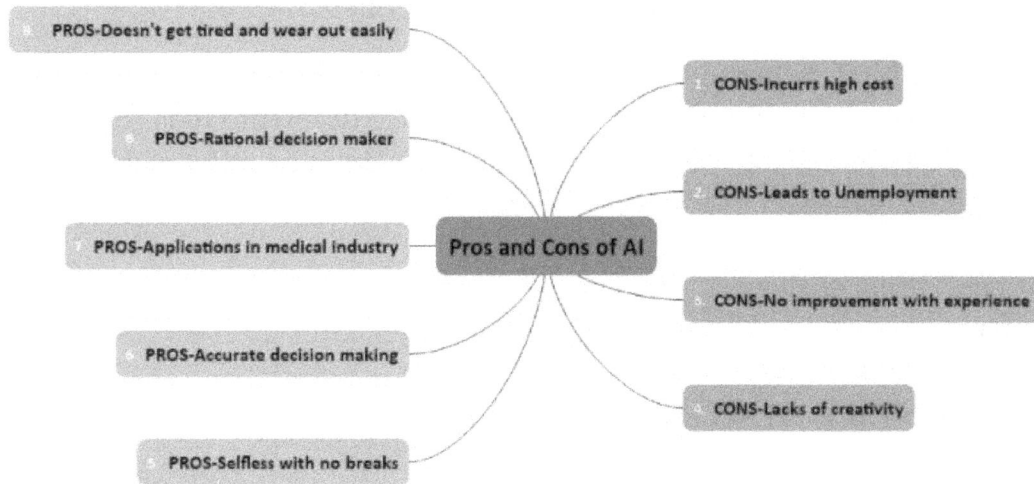

Figure 8.7 Pros and cons of artificial intelligence.

8.6 CHALLENGES

Just like other technologies, AI comes with challenges, such as accountability, security, technological mistrust, and the displacement of human workers. These are clear challenges that must be addressed to support AI technology's future. Digitalization and cyber security should go hand in hand. The risks are huge without the right safeguards in place. Standards of AI should be created to match the perspectives of societal acceptance. The major challenges of industrial AI include [2]:

• **Data**: Modern industry is indeed a big data environment. Industrial data usually is structured but may be of low-quality.

• **Speed**: Production process happens fast, and the equipment and work piece can be expensive.

• **Fidelity**: A very low rate of false positives or negatives rate may cost the total credibility of AI systems. Industrial AI applications are usually dealing with critical issues related to safety, reliability, and operations.

- **Interpretability:** The industrial AI systems must go beyond prediction results and give root cause analysis for anomalies.

- **Scaling challenges:** It is difficult to integrate AI projects into existing system. The process for using AI takes too long or is expensive to scale. In scaling up, companies may face substantial change-management challenges.

- **Hammer/Nail Scheme:** When an AI company has a hammer, everyone else's problem starts to look like a nail. Within industries, different kinds of expertise will be critical to a successful application of AI. These fundamental differences in each industry should not be ignored.

Regardless of the challenges, AI is posed to affect the world's economies, citizens, and the Internet.

8.7 INDUSTRY ACROSS THE GLOBE

Harnessing the power of AI is fast becoming a top global concern for businesses since AI is world-changing technology, transforming the industry across the globe. AI offers tremendous potential for industry. AI technologies worldwide have gained significant momentum over the past decade and become more mainstream. They are emulating human cognition at increasingly higher levels of abstraction and adaptation capabilities.

- **United States:** In the US, AI has been the focus of research for more than 30 years. Recently, to accelerate leadership in AI initiative, the US government launched an official website AI.gov to highlight its priorities in the AI space. NIST, along with external partners, is developing testing methods and metrics to help industry better pick out useful AI tools from "bad" candidates. Only in the US, 57 chains, including Sears and Toys'R'Us, filed Chapter 11 bankruptcy protection in 2015. From driverless cars to virtual doctors, AI is transforming the way we live, work, travel, and do business in the 21st century.

- **China:** The AI industry has developed rapidly in China in recent years. China's governments have set policies to support this industry. However, China is facing challenges of lacking industry standards, data desensitization standard, assessment system, and policies to realize the AI products.

The Chinese Innovative Alliance of Industry, Education, Research and Application of Artificial Intelligence for Medical Imaging (CAIERA), the first Chinese organization sponsored by the demand-apply was formed in April 2018 and compiled the White Paper on Medical Image AI in China. The white paper calls on people from all stakeholders pay active attention to the development of medical imaging AI. The alliance has issued a certain number of consensuses on several issues in AI. Development of AI in medical imaging requires breakthroughs of the core algorithm, commitment of medical practitioners, and the resolutions from government [22]. The development of AI in China has important implications for other emerging markets. Figure 8.8 shows China AI investment by industrial subsector [23].

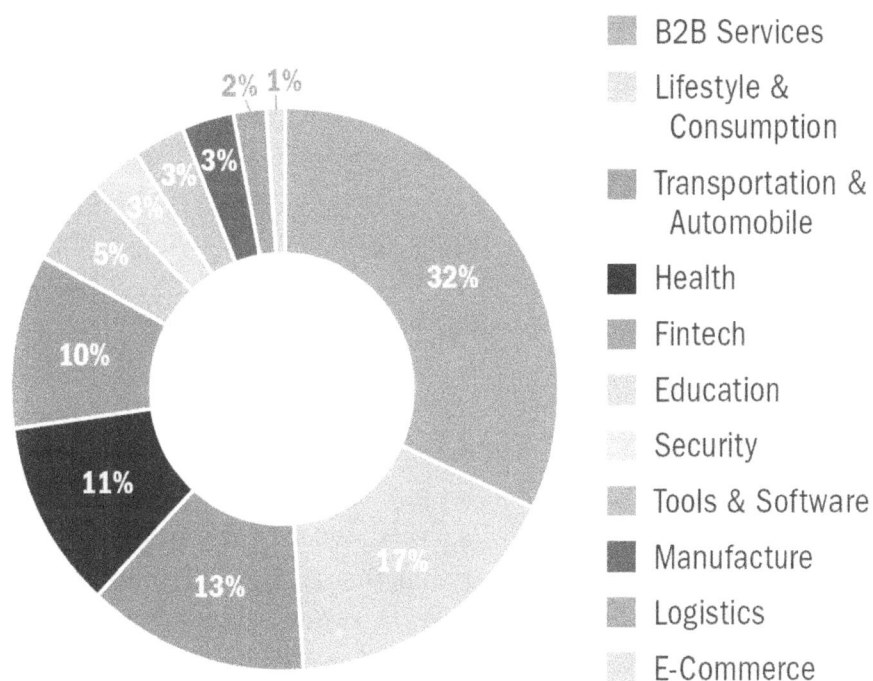

Figure 8.8   China AI investment by industrial subsector [23].

•   **United Kingdom:** Increased use of AI can bring major social and economic benefits to the UK. AI offers massive gains in efficiency and performance to most or all industry sectors. The UK aims at being among the leaders in AI and to become the best place in the world for businesses development. While other nations are investing heavily in AI development, the UK is regarded as a center of expertise, for the moment at least. To continue developing and applying AI, the UK will need to increase ease of access to data in a wider range of sectors. Skilled experts are needed to develop AI, and they are in short supply. To develop more AI, the UK will need a larger workforce with deep AI expertise [24].

• **European Union:** Business opportunities for European companies include AI technologies and AI solutions. The European AI policy landscape must ensure legal certainty while providing a dynamic ecosystem capable of adapting to the constant evolution of technology. Europe's automobile manufacturers understand that it is an absolute priority for them to put vehicles on the market that meet the maximum safety and ethical standards. Mandatory requirements should be reserved exclusively for high-risk, safety-critical AI applications related to automated driving. In the future, AI-based solutions can be deployed for map building, image analysis, and data fusion [25].

• **Japan:** The Japanese government is betting on AI as the key to rewrite Japan blueprint for the future. However, many Japanese companies have been relatively slow in adopting AI technologies. Some Japanese companies have sought outside partners to further their ambitions. The automobile industry in Japan will need AI solutions to develop driverless cars. Several headlines in recent times have brought to the attention the current status of the Japanese AI industry. Japan is still at the forefront in hardware such as robots, but it lags in software, where it has lost its technology leadership to companies in the west. The merger of AI applications and consumer products will increase in the future. Businesses in many sectors of the economy are now using AI technologies/solutions [26].

8.8 FUTURE OF INDUSTRIAL AI

AI has a rapidly growing presence in today's world, with applications ranging from heavy industry to education. It is becoming clear that this technology can revolutionize how the everyday world works. AI technology will contribute greatly to the increase in global economic growth and productivity. If you have interest in how developments in AI might impact your business, you must keep an eye on trends of industrial AI.
As AI emerges from science fiction to become the frontier of world-changing technologies, there is an urgent need for standards and systematic development and implementation of AI. Businesses lose a lot through cybercrime as hackers are growing smarter with time. Comprehensive protection for industrial facilities will play a crucial role in the future.

8.9 CONCLUSION

Artificial intelligence has gradually become well integrated into many aspects of society and widely adopted in the industry. AI technology can change the world for the better. The disruptive achievements of AI are rapidly infiltrating into various fields of human activities. AI is getting more sophisticated with time and its programs are revolutionizing the way that some industries operate. Google Maps, LinkedIn, Facebook, Uber, and many other future-looking companies are using AI to improve their services. More information about industrial AI can be found in the books in [27-29].

REFERENCES

[1] "Artificial intelligence in industry: Intelligent production,"
https://new.siemens.com/global/en/company/stories/industry/ai-in-industries.html#:~:text=Artificial%20intelligence%20offers%20tremendous%20potential,more%20flexible%2C%20and%20more%20reliable.&text=The%20volumes%20of%20data%20in,plants%20and%20systems%20are%20generated.

[2] "Artificial intelligence in industry," Wikipedia, the free encyclopedia
https://en.wikipedia.org/wiki/Artificial_intelligence_in_industry

[3] **M. N. O. Sadiku, S. M. Musa**, and **O. M. Musa**," The essence of Industry 4.0," Invention Journal of Research Technology in Engineering and Management, vol. 2, no. 9, September 2018, pp. 64-67.

[4] **H. Chen, L. Li**, and **Y. Chen**, "Explore success factors that impact artificial intelligence adoption on telecom industry in China," Journal of Management Analytics, 2020.

[5] **M. N. O. Sadiku**, "Artificial intelligence", IEEE Potentials, May 1989, pp. 35-39.

[6] **S. Greengard**, "What is artificial intelligence?" May 2019,
https://www.datamation.com/artificial-intelligence/what-is-artificial-intelligence.html

[7] https://www.researchgate.net/figure/Branches-of-Artificial-Intelligence-AI_fig2_322113260

[8] **M. Sharp**, "A.I. for smarter factories: The world of industrial artificial intelligence," September 2020,
https://www.nist.gov/blogs/taking-measure/ai-smarter-factories-world-industrial-artificial-intelligence

[9] **J. Lee et al.,** "Industrial artificial intelligence for industry 4.0-based manufacturing systems," Manufacturing Letters, vol. 18, October 2018, pp. 20-23.

[10] **S. Charrington,** "Artificial intelligence for industrial applications," https://uploads-ssl.webflow.com/5a6a107a0a6e6500019f9a5d/5bd87cc016e1ea3efb5c4420_CloudPulse_Industrial_AI_Report_2017062301.pdf

[11] **D. W. Hall** and **J. Pesenti**, "Growing the artificial intelligence industry in the UK," https://assets.publishing.service.gov.uk/government/uploads/system/uploads/attachment_data/file/652097/Growing_the_artificial_intelligence_industry_in_the_UK.pdf

[12] **The European Automobile Manufacturers' Association (ACEA)**, "13 Industries soon to be revolutionized by artificial intelligence," January 2019, https://www.forbes.com/sites/forbestechcouncil/2019/01/16/13-industries-soon-to-be-revolutionized-by-artificial-intelligence/?sh=242823b63dc1

[13] **V. Vikram**, "12 Industries artificial intelligence will revolutionize," November 2019, https://www.icicletech.com/blog/industries-artificial-intelligence-impacts

[14] **E. Clark**, "3 Industries most affected by artificial intelligence," January 2020, https://fowmedia.com/3-industries-most-affected-by-artificial-intelligence/

[15] **R. Day**, "Manufacturing enters era of artificial intelligence," August 2017, https://www.business.uconn.edu/2017/08/14/manufacturing-enters-era-of-artificial-intelligence/#

[16] **R. Reynoso**, "7 Industries using AI in 2020 (+14 Examples)," June 2019, https://learn.g2.com/industries-using-ai

[17] **N. Achary**, "Artificial intelligence to transform 10 Industries," March 2019, https://becominghuman.ai/artificial-intelligence-to-transform-10-industries-498338359f41

[18] **J. Qi et al.**, "Artificial intelligence applications in the telecommunications industry," *Expert Systems,* vol. 24, no. 4, September 2007, pp. 271-291.

[19] **M. N. O. Sadiku, S. M. Musa**, and **O. S. Musa**," Artificial intelligence in chemical industry," International Research Journal of Advanced Research in Science, Engineering and Technology, vol. 4, no. 10, Oct. 2017, pp. 4618-4620.

[20] **A. Takyar**, "AI applications across major industries," HTTPS://WWW.LEEWAYHERTZ.COM/AI-APPLICATIONS-ACROSS-MAJOR-INDUSTRIES/

[21] "Pros and cons of artificial intelligence – A threat or a blessing?" https://data-flair.training/blogs/artificial-intelligence-advantages-disadvantages/

[22] **Y. Xiao** and **S. Liu**, "Collaborations of industry, academia, research and application improve the healthy development of medical imaging artificial intelligence industry in China," Chinese Medical Sciences Journal, vol. 34, no. 2, 2019, pp. 84-88.

[23] **X. Mou**, "Artificial intelligence: Investment trends and selected industry uses," https://www.ifc.org/wps/wcm/connect/7898d957-69b5-4727-9226-277e8ae28711/EMCompass-Note-71-AI-Investment-Trends.pdf?MOD=AJPERES&CVID=mR5Jvd6

[24] **D. W. Hall** and **J. Pesenti**, "Growing the artificial intelligence industry in the UK," https://assets.publishing.service.gov.uk/government/uploads/system/uploads/attachment_data/file/652097/Growing_the_artificial_intelligence_industry_in_the_UK.pdf

[25] The European Automobile Manufacturers' Association (ACEA), "Artificial Intelligence in the automobile industry," https://www.acea.be/uploads/publications/ACEA_Position_Paper-Artificial_Intelligence_in_the_automotive_industry.pdf

[26] "Artificial intelligence in Japan (R&D, market and industry analysis) https://www.eubusinessinjapan.eu/sites/default/files/artificial_intelligence_in_japan.pdf

[27] **A. Azizi**, Applications of Artificial Intelligence Techniques in Industry 4.0. Springer, 2019.

[28] **E. G. Popkova, J. V. Ragulina**, and **V. Aleksei (eds.)**, Industry 4.0: Industrial Revolution of the 21st Century. Springer, 2019.

[29] **A. Dingli, F. Haddod**, and **C. Klüver (eds.)**, Artificial Intelligence in Industry 4.0. Springer, 2021.

CHAPTER 9
AI IN MANUFACTURING

"Technology, through automation and artificial intelligence, is definitely one of the most disruptive sources." -*Alain Dehaze*.

9.1 INTRODUCTION

Manufacturing refers to the entire product life cycle: product design, production planning, production, distribution, and field service and reclamation [1]. The manufacturing industry is a cornerstone of national economy, and people's livelihood. The industry has always been receptive to adopting new technologies such as drones, industrial robots, artificial intelligence (AI), virtual reality, and Internet of things. These days, humans and robots collaborate to produce breakthrough technologies.

Traditional manufacturing companies currently face several challenges such as rapid technological changes, inventory problem, shortened innovation, short product life cycles, volatile demand, low prices, highly customized products, and ability to compete in the global markets. Today, the manufacturing industry is highly competitive due to globalization and fast changes in the global market. Manufacturing systems are evolving into a new generation of data-driven systems that must adapt to these changes in a timely fashion. Using artificial intelligence and machine learning to streamline every phase of production and increase productivity is now a priority in manufacturing.

Artificial intelligence (AI) is the cognitive science that deals with intelligent machines which are able to perform tasks heretofore only performed by human beings. It is mainly concerned with applying computers to tasks that require knowledge, perception, reasoning, understanding, and cognitive abilities. AI is potentially the algorithmic study of processes in every field of study [2]. The main objective of AI is to teach the machines to think intelligently like humans do.

Artificial Intelligence began to become an active field of research within computer science in about 1955. Today, AI in smart machines handles the more traditional repetitive tasks. AI has many applications in today's society. Although AI is a branch of computer science, there is hardly any field which is unaffected by this technology.

Common areas of applications include agriculture, business, law enforcement, oil and gas, banking and finance, education, transportation, healthcare, automobiles, entertainment, manufacturing, speech and text recognition, facial analysis, and telecommunications. Over the years, the field of AI has produced a number of tools for manufacturing. Today industrial leaders such as Google, Microsoft, Procter & Gamble, and IBM have invested heavily on AI [3].

This chapter briefly addresses the uses of artificial intelligence in the manufacturing industry. It begins by briefly reviewing AI. It discusses the roles AI plays in manufacturing. It highlights the benefits and challenges of applying AI in manufacturing. It covers how AI is being applied in manufacturing around the world. It addresses the future of AI in manufacturing. The last section concludes with comments.

9.2 REVIEW ON ARTIFICIAL INTELLIGENCE

In simple terms, artificial intelligence (AI) refers to computer systems that mimic human cognitive functions. The term "artificial intelligence" (AI) was first used at a Dartmouth College conference in 1956. The main goal of AI is to enable machines to perform complex tasks that typically require human intelligence [4]. AI is now one of the most important global issues of the 21st century.

AI is not a single technology but a range of computational models and algorithms. AI is a collection of techniques that enables computer systems to perform tasks that would otherwise require human intelligence [5]. The major disciplines in AI include:

- Expert systems
- Fuzzy logic
- Neural networks
- Machine learning

- Deep learning
- Natural Language Processors
- Robots

These AI tools are illustrated in Figure 9.1.

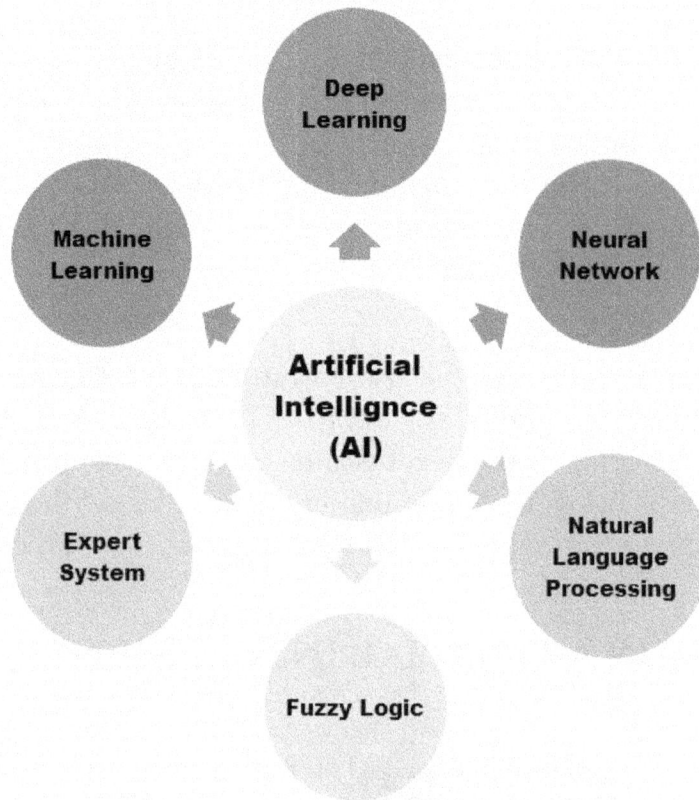

Figure 9.1 Branches of artificial intelligence.

Each AI tool has its own advantages. Using a combination of these models, rather than a single model, is recommended. AI systems are designed to make decisions using real-time data. They have the ability to learn and adapt as they make decisions. Artificial intelligence is no longer just an academic field; machine learning and deep learning are becoming mainstream technologies that manufacturing companies can harness.

AI-driven systems can discover trends, reveal inefficiencies, and predict future outcomes. These characteristics enable informed decision-making and make AI to be potentially beneficial for many industries. Several companies have realized the value of AI technology. By building precise models, a company has a better chance of identifying profitable opportunities and avoiding risks. Machine learning techniques are being applied in several sectors including government, healthcare, finance, retail, transportation, oil and gas, and manufacturing [6]. The factors that make AI best suitable for manufacturing include [7]: virtual reality, automation, intelligence, prediction, and better products.

9.3 AI IN MANUFACTURING

Manufacturing is regarded as a prime generator of wealth and is critical in establishing a sound basis for economic growth. The leading manufacturing industries in US include steel, automobiles, chemicals, food processing, consumer goods, aerospace, and mining. Globalization has radically changed the manufacturing process in recent years. The manufacturing industry has become more competitive due to globalization. To survive from the global market, manufacturing enterprises should reduce the product cost and increase the productivity.

Manufacturing consists of a set of processes, machines, and factories where raw materials are transformed into products. The search for new ways to increase efficiency and productivity has always been at the heart of manufacturing. To meet these challenges, it is essential to utilize all means available. One area, which saw fast-paced developments in terms of not only promising results but also usability, is artificial intelligence. AI is one of the many technologies currently impacting manufacturing. Over the years, the manufacturing industry has exploited the use of artificial intelligence (AI), which is the intelligence exhibited by machines [8].

Manufacturing efficiency and maintaining product quality remain top priorities in some industries such as integrated circuits (IC) industry, where process and equipment reliability directly influence cost, throughput, and yield [9]. Recent developments have enabled AI to cross into the mainstream: cloud computing, big data, machine learning, and Internet of things (IoT).

The manufacturing industry is now moving into the fourth revolution [10]:

- The 1st revolution introduced mechanization through water and steam power.
- The 2nd revolution ushered in mass production and assembly lines using electricity.
- The 3rd industrial revolution introduced the digital era, comprised of computers and automation.
- The 4th industrial revolution, referred to as "Industry 4.0," is happening now.

These revolutions are illustrated in Figure 9.2.

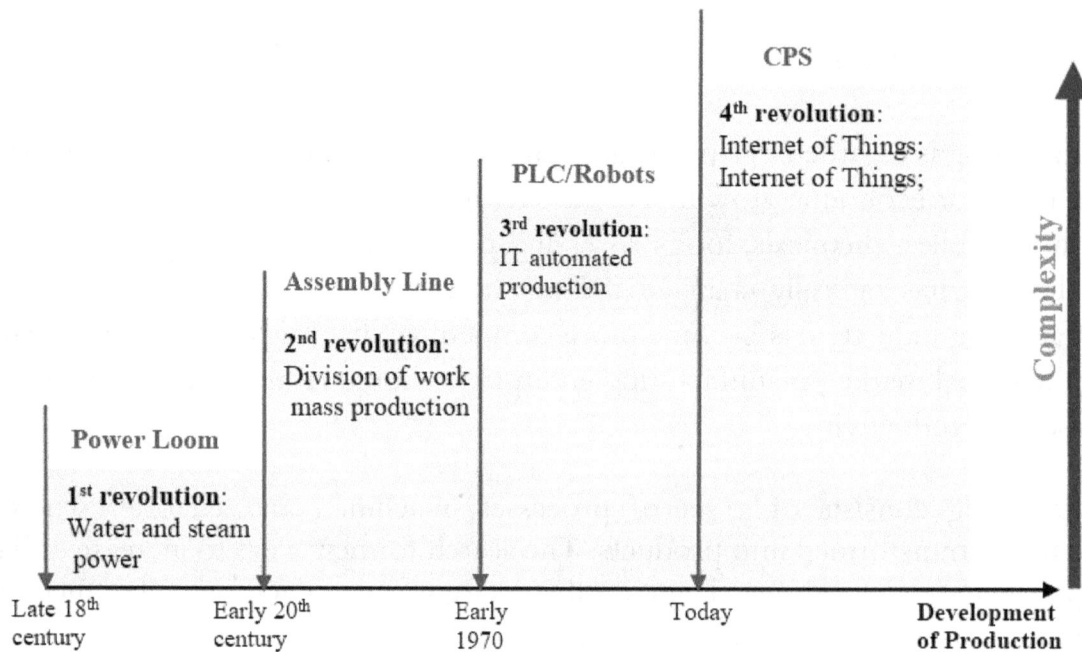

Figure 9.2 Four industrial revolutions.

9.4 APPLICATIONS OF AI IN MANUFACTURING

The techniques of the artificial intelligence (IA) in general and machine learning (ML) have been a source of inspiration for researchers in the manufacturing industry.

Robotics: Perhaps the most promising way is applying machine learning algorithms to the traditional manufacturing system. Even with advanced manufacturing techniques, using humans to spot defects and errors is very limiting. Today, robotics has revolutionized manufacturing, allowing for greater output from fewer workers, while AI is just beginning to live up to its full potential. A robot is a mechatronic device that is designed and programmed to perform some specific tasks. Robots typically use sensor data to make decisions. Robots are used in diverse applications such as pick and place, sorting, assembly, painting, welding, storage, and retrieval, AI coupled with computer vision techniques allows autonomous robots to complete these tasks. A typical example of the use of robots in manufacturing is shown in Figure 9.3 [11].

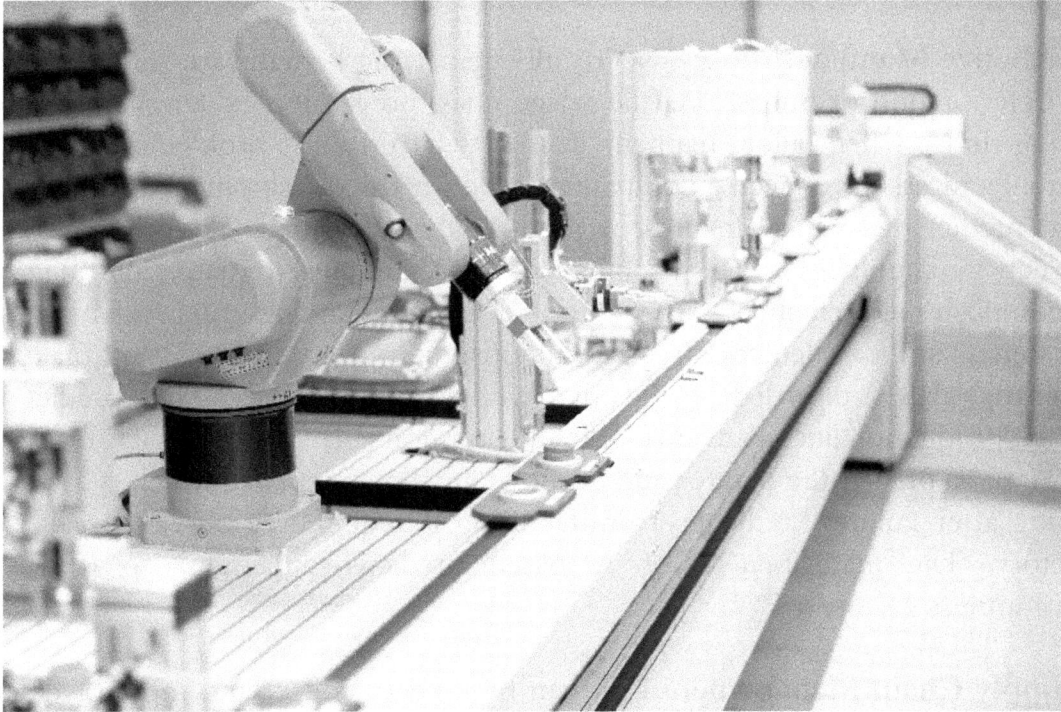

Figure 9.3 A example of the use of robots in manufacturing [11].

Industry 4.0: The manufacturing industry today is experiencing an unprecedent increase in available data. These data that can be obtained from manufacturing system consists of man, machine, material, and method data (i.e., 4M data). Different names are used for this phenomenon, e.g., Industry 4.0 (Germany), Smart Manufacturing (USA), and Smart Factory (South Korea). Industry 4.0 refers to the current trend of automation and employment of AI technologies in manufacturing. The major applications of Industry 4.0 are smart Factory, manufacturing, smart product, and smart city. Those who promote Industry 4.0 claim that it will affect many areas such as services and business models, productivity, machine safety, product lifecycles, and industry value chain [12,13]. The terms "smart factory," "smart manufacturing," "industry 4.0" and "factory of the future" represent keywords to denote what industrial production will look like in the future. They refer to a factory with a manufacturing solution that provides a flexible, adaptive, and efficient production. The implementation of the smart factory is enabled by several technologies such as artificial intelligence, industrial robots, embedded systems, RFID, and the Internet of things (IoT) or industrial Internet of things (IioT).

Predictive Manufacturing: Nearly all engineering systems are subject to failures due to deterioration or misuse. As the adage goes, prevention better than cure, instead of fixing things when failure happens, it is better to predict problems before they occur. Manufacturers have faced a compelling need for the development of predictive models that predict mechanical failures or predict when maintenance is due. The one main objective of applying AI and machine learning techniques to manufacturing systems is to accomplish the predictive manufacturing. This is a manufacturing system that has a self-awareness and can detect current conditions, diagnose a problem, and make a decision. This may also be regarded as predictive maintenance, which is able to predict disruptions to the production line in advance of that disruption [14]. Data collected from sensors can be used to predict the failures and do preventive maintenance which is lot cheaper than fixing after failure. Predictive manufacturing is required in order to realize the smart factory. The main benefit of AI-based predictive maintenance is accuracy and promptness.

Supply Chain Management: AI can be used to collect and monitor data along the supply chain, and then manage inventory, predict future demand, and detect inefficiencies. Artificial intelligence and machine learning make supply chain management not only automated but cognitive. The cognitive supply chain ensures data privacy and prevents hacking. Supply chain management systems based on AI algorithms can automatically analyze such data, define optimal solutions, and make data-driven decisions. Supply chain management is an example of how AI and big data can be used to the benefit of manufacturers.

Intelligent Manufacturing: The continuous evolution of smart cities, intelligent transportation, intelligent robots, smart-phones, smart communities, and smart grid, etc. provides a market demand and driving force for both AI technologies and applications. This has led to a new manufacturing model known as intelligent or smart manufacturing. AI is applied in intelligent manufacturing through the intelligent manufacturing system. The intelligent manufacturing system is characterized by autonomous intelligent sensing, intelligent equipment, intelligent robots, new-generation intelligent network devices, networking, collaboration, learning, analysis, cognition, knowledge-based intelligence design, prediction, decision-making, and on-demand services at any time and any place. Intelligent manufacturing system is used to facilitate production and provide a high efficiency, high quality, cost-effective, and environment-friendly service to users [15]. Figure 9.4 shows the components of an intelligent manufacturing system.

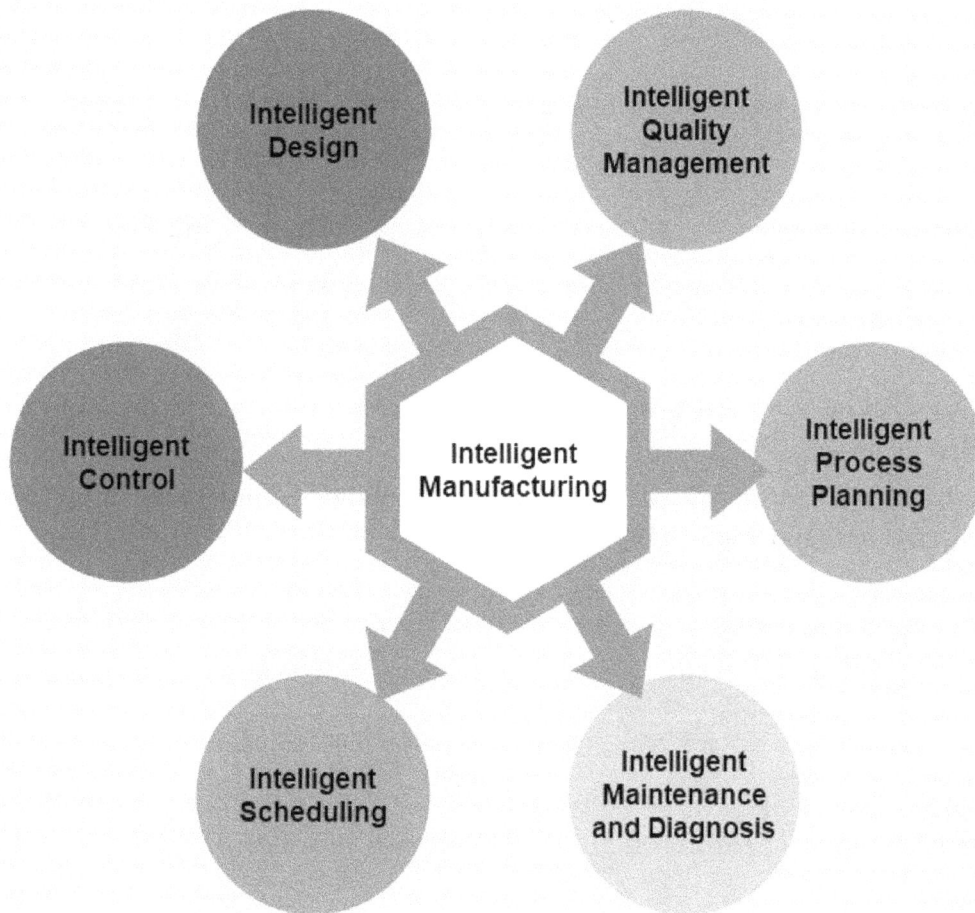

Figure 9.4 Components of an intelligent manufacturing system.

Artificial intelligence is also used in additive manufacturing, digital manufacturing, smart manufacturing, and generative design.

9.5 BENEFITS

Artificial intelligence can help address some issues facing manufacturing, improve quality control, shorten design cycles, remove supply-chain bottlenecks, reduce materials and energy waste, and improve production yields. AI tools are helping manufacturers find new business models, improve their product quality, increase productivity, reduce chronic labor shortage, and optimize manufacturing operations, while being able to take on short lead-time production runs from customers. They can enhance and extend the capabilities of humans. Other benefits include [16,17]:

• **Robotics**: A large share of the manufacturing jobs is performed by robots. AI-powered robots can interpret CAD models, which eliminates the need to program their movements and processes. Industrial robots perform what they were programmed to do, but they cannot make complex decisions as humans.

• **Higher Productivity**: With the help of AI, manufacturing will lead to higher productivity, increase efficiency, and improve the way customers are served. AI is poised to impact the supply chain.

• **Better Decisions**: With the adoption of AI, manufacturers are able to make rapid, data-driven decisions, optimize manufacturing processes, minimize operational costs. While machine cannot yet surpass human intelligence, they can outperform humans in sheer speed, accuracy, and ability to search vast amounts of information.

• **Price Forecasts**: Knowing the prices of resources is also necessary for companies when their product is ready to leave the factory. The system is able to provide accurate price recommendations.

• **Customer Service**: AI solutions can analyze the behaviors of customers to identify patterns and predict future outcomes. There is a number of AI tools that can improve customer service.

• **Preventative Maintenance**: Machine learning can be used to predict preventative maintenance in order to improve worker safety, reduce costs, and achieve sustainability goals.

9.6 CHALLENGES

The field of AI is broad and even confusing, presenting a challenge and hindering wide application. Manufacturing is a high-cost venture for most businesses, and it requires considerable capital investments. It is possible for AI to develop a will of its own that may be in conflict with ours. The increasing use of connected technologies makes the manufacturing system vulnerable to cyber-attacks [18]. Other challenges include [17].

- **Complexity:** A major challenge is the ongoing trend of the manufacturing domain to becoming more complex and dynamic. Due to this, modern manufacturing technology is interdisciplinary in nature.

- **Common Fear:** There is the common fear of human jobs being lost to AI. A good example is a single associate oversees about 10 self-checkout point-of-sale (POS) machines at Walmart. Mass unemployment will be a social challenge. This will require retraining human workers for new careers. With automation, it is easy to believe that machines outperform human workers at every turn. Due to this imminent trend, businesses should train their workers to work by the side of AI and come to grips with the impending obligation.

- **Environmental Impact:** The manufacture of a variety of products continues to damage the environment.

- **Shortage in Workforce:** Manufacturing companies will feel the challenge of a decreasing talent pool. There are not enough skilled people to perform the AI-related jobs of the future. There continues to be a shortage within the manufacturing workforce.

- **Liability:** AI raises important legal, ethical, and public policy questions. AI will have to make potentially life-or-death decisions. How should a self-driving car decide between crashing itself and potentially killing its passengers? How do we assign liability for when AI makes mistakes?

9.7 GLOBAL AI IN MANUFACTURING

Artificial intelligence is gradually being implemented in almost every nation. In recent years, the global market has witnessed significant progress in technologies, industry, and applications in terms of manufacturing in general and intelligent manufacturing in particular. With the inroading of Artificial Intelligence (AI) and Internet of Things (IoT) into the manufacturing sector, nations are inevitably faced with a number of policy concerns which need to be addressed at various levels keeping in mind the socio-economic factors that influence policy making in that particular nation. For example, the United States and Germany have spared no effort in demonstrating and promoting their development strategy to transform and upgrade their manufacturing industries.

The Global Manufacturing and Industrialization Summit (GMIS) was established in 2015 to build bridges between manufacturers, governments and NGOs, technologists, and investors so that they can harness the transformative power of the Fourth Industrial Revolution. Here we consider how AI is being applied in manufacturing in different countries [13,19].

United States: Intelligent manufacturing technology in the US has accomplished initial outcomes. Boeing Company and General Electric have identified challenges to the incorporation of AI in the manufacturing sector. The Manufacturing Institute and Deloitte estimates that the United States alone may have as many as 2.4 million manufacturing jobs to fill between now and 2028, Americans have not yet grappled with just how profoundly the artificial intelligence (AI) revolution will impact our economy, national security, and welfare. We take seriously China's ambition to surpass the United States as the world's AI leader within a decade.

China: The manufacturing industry in China is facing a critical and historical moment in terms of the transformation manufacturing giant to manufacturing power and from "made in China" to "created in China." The industry is in a phase of imbalanced development among different regions, industries, and enterprises, where mechanization, electrification, automation, and digitization co-exist. The Chinese government has proposed strategic plans of "made in China 2025." It sets the principles for becoming the second tier of the powerhouse in manufacturing: innovation-driven, quality first, environment-friendly development, structural optimization, and talent-centric. The plan also includes the following nine missions: (1) increasing innovative capability in national manufacturing, (2) promoting the deep fusion of information and industrialization, (3) strengthening the basic industry capacity, (4) booming the quality brand-building, (5) popularizing environment-friendly manufacturing, (6) advancing breakthroughs in key areas, (7) pushing forward further the structural adjustment to the manufacturing industry, (8) advancing services for manufacturing and production, and (9) increasing international involvement in manufacturing.

Germany: Germany mainly focuses on the research of underlying technologies for manufacturers, such as intelligent sensing, wireless sensor networks, and CPSs. Sensors will gather data from the engine during its flight and transmit it to the ground for smart analysis. The Germany Amberg factory is a model of an intelligent plant of the Siemens company. At Amberg, the real factory is operated together with the virtual factory, and the real factory data and production environments are reflected by the virtual factory.

India: The impact of AI in the manufacturing in India often depict a picture of stagnant job growth and even job loss. AI is disrupting traditional business models in the IT sector, the automotive industry, and other manufacturing industries. However, smart manufacturing is picking up in India. The Indian Institute of Science is developing a smart factory. India, unlike its G20 counterparts, is yet to fully tap the available opportunities that AI presents.

9.8 FUTURE OF AI IN MANUFACTURING

Global competition and continuously changing customer requirements are dictating increasing changes in manufacturing environments. AI in manufacturing is a game-changer. AI is already transforming manufacturing in several ways and also changing the way we design products. Today manufacturing has a strong association with AI, especially the use of automation. AI technologies in the manufacturing industry are expected to grow over the next years. There is an excellent future for use of AI in manufacturing due to the potentially high value added when applied successfully [20]. The future will witness a deep fusion of new AI technologies with Internet technologies. AI will have a fundamental impact on the global labor market in the next few years.

Today's technologies are able to predict design success, product performance, and equipment failures, providing key information that helps manufacturers continuously improve. AI algorithms formulate estimations of market demands by looking for patterns linking location, socioeconomic and macroeconomic factors, weather patterns, political status, consumer behavior, and more. This information is invaluable to manufacturers as it allows them to optimize staffing, inventory control, energy consumption, and the supply of raw materials.

Some experts in the field of manufacturing are already making predictions on what is coming and the trends driving AI advancement. Figure 9.5 illustrates the future of AI in manufacturing. Here are three exciting trends to watch out for coming years [21]:

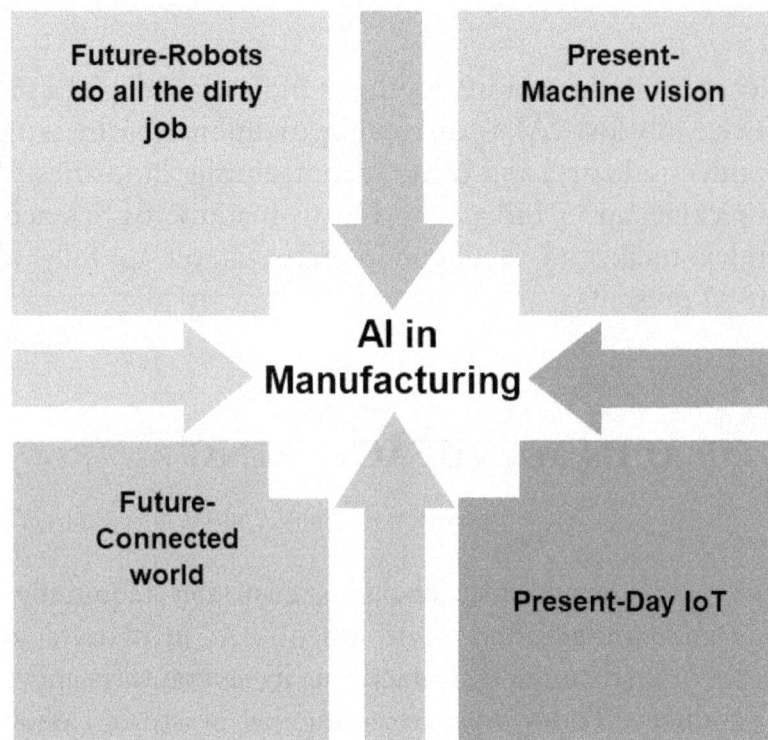

Figure 9.5 The future of AI in manufacturing.

Advanced Metering Infrastructure (AMI): This is a family of activities that depend on the use and coordination of information, automation, computation, software, sensing, and networking. It uses cutting edge materials and emerging technologies such as nanotechnology and industrial IoT. In order to capture the potential of advanced manufacturing and create new advantages, we must collaborate to bring these emerging technologies into the next generation. With artificial intelligence, advanced manufacturing can become more autonomous, efficient, and profitable [22].

Smart Manufacturing: Smart manufacturing (SM) is a term coined by several agencies in the United States and increasingly used globally. Other initiatives such as Industry 4.0, cyber-physical production systems, smart factory, intelligent manufacturing, and advanced manufacturing are frequently used synonymously with smart manufacturing. SM is characterized by the integration of new-generation information communication technology (ICT) and manufacturing industry. As shown in Figure 9.6, smart manufacturing, which is the fourth revolution in the manufacturing, has its own identity captured in the following six pillars.

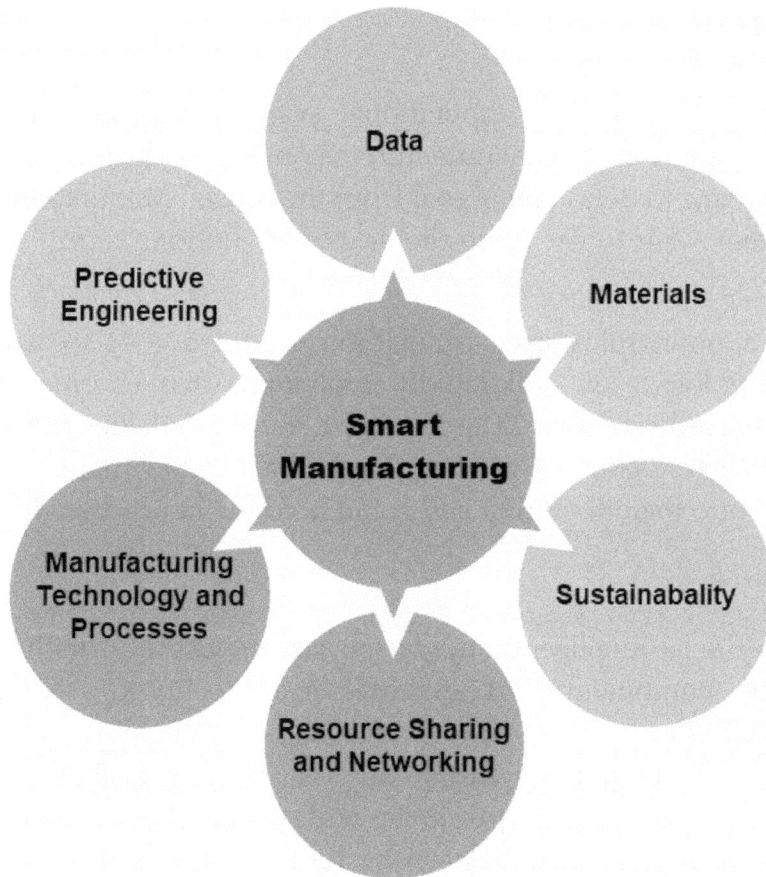

Figure 9.6 Six pillars of smart manufacturing.

Smart manufacturing will transform how products are designed, fabricated, operated, and used. Globalization and competitiveness are forcing companies to re-think and innovate their production process [23,24].

Additive Manufacturing: Additive manufacturing (AM) (or 3D-printing refers to a process by which 3D CAD data is used to construct an object in layers by depositing material. The promise of AM is the use of less material and to provide faster cycle time due to rapid prototyping. AM technologies typically include a computer, 3D modeling software (Computer Aided Design or CAD), a machine and material. The AM machine reads in data from the CAD file and develops successive layers of the material in a layer-upon-layer manner to fabricate the desired 3D object. Additive manufacturing allows one to produce the objects without machining, lathing, milling, grinding, boring, casting or welding. It will take manufacturing to the next level. It will lead to home manufacturing [25].

9.9 CONCLUSION

Artificial intelligence deals with computational systems that imitate the intelligent behavior of human expertise. For manufacturers considering introducing AI into their processes, it is important to define their goal from the outset. Manufacturers need to ask themselves where they want to use AI in the near, mid and long term.

AI is both a powerful source of disruption and a tool to gain competitive advantage. AI technologies are still rapidly evolving, often causing uncertainty for companies around what their future IT architecture should look like. For manufacturers, AI promises to be a game-changer at every level of the value chain. AI technology is now making its way into manufacturing and could hold the key to transforming factories of the near future.

Without doubt artificial intelligence holds the key to future growth and success in manufacturing. The combination of manufacturing technology with information communication technology and smart technology is enabling a game-changing transformation in terms of manufacturing models and approaches. AI is poised to change the way we manufacture products and process materials forever. More information about AI in manufacturing can be found in the books in [26,27] and the following related journals:

- Applied Artificial Intelligence
- Artificial Intelligence for Engineering Design, Analysis and Manufacturing.
- IEEE Journal on Robotics and Automation
- Journal of Intelligent Manufacturing
- AI Magazine
- Manufacturing Letters

REFERENCES

[1] **M. S. Fox**," Industrial applications of artificial intelligence," Robotics, vol. 2, 1986, pp. 301-311.

[2] **M. N. O. Sadiku**, "Artificial intelligence", IEEE Potentials, May 1989, pp. 35-39.

[3] **M. N. O. Sadiku, S. M. Musa, and O. M. Musa**, "Artificial intelligence in the manufacturing industry," International Journal of Advances in Scientific Research and Engineering, vol. 5, no. 6, June 2019, pp. 108-110.

[4] **H. Chen, L. Li**, and **Y. Chen**, "Explore success factors that impact artificial intelligence adoption on telecom industry in China," Journal of Management Analytics, 2020.

[5] **S. Greengard**, "What is artificial intelligence?" May 2019,
https://www.datamation.com/artificial-intelligence/what-is-artificial-intelligence.html

[6] "How does artificial intelligence work?"
https://www.innoplexus.com/blog/how-artificial-intelligence-works/

[7] **T. Nicholas**, "The future of artificial intelligence in manufacturing industries,"
December 2018,
https://www.intellectyx.com/blog/the-future-of-artificial-intelligence-in-manufacturing-industries/

[8] **M. N. O. Sadiku**, Y. **P. Achara**, and **S. M. Musa**, "Machine learning in manufacturing: A brief survey," International Journal of Trend in Research and Development, vol. 6, no. 6, December 2019.

[9] "Artificial intelligence in semiconductor manufacturing," in J. Webster (ed.), Wiley Encyclopedia of Electrical and Electronics Engineering, John Wiley & Sons,2007.

[10] P. **P. Khargonekar**, "Artificial intelligence, automation, and manufacturing,"
June 2018.
https://cpb-us-e2.wpmucdn.com/faculty.sites.uci.edu/dist/8/644/files/2018/06/Smart_MFG_CNMI.pdf

[11] **B. Ramsey**, "How machine learning is poised to revolutionize manufacturing,"

https://medium.com/supplyframe-hardware/how-machine-learning-is-poised-to-revolutionize-manufacturing-7e72b4ba8e5f

[12] **T. West et al.**, "Machine learning in manufacturing: Advantages, challenges, and applications," Production & Manufacturing Research, vol. 4, no.1, 2016, pp. 23-45.

[13] **M. N. O. Sadiku, S. M. Musa,** and **O. M. Musa**," The essence of Industry 4.0," Invention Journal of Research Technology in Engineering and Management, vol. 2, no. 9, September 2018, pp. 64-67.

[14] **J. H. Han, R. Kim,** and **S. Y. Chi**, "Applications of machine learning algorithms to predictive manufacturing: Trends and application of tool wear compensation parameter recommendation," *Proceedings of the 2015* International Conference on Big Data Applications and Services, Jeju Island, Korea, October 2015, pp. 51-57.

[15] **B. Li et al.**, Applications of artificial intelligence in intelligent manufacturing: a review," Frontiers of Information Technology & Electronic Engineering, vol.18, February 2017, pp. 86–96.

[16] **F. Meziane et al.**, "Intelligent systems in manufacturing: Current developments and future prospects," https://www.researchgate.net/publication/242176196_Intelligent_systems_in_manufacturing_Current_developments_and_future_prospects/link/02bfe5110ffd9b3797000000/download

[17] **K. Polychasia**, "10 use cases of AI in manufacturing," June 2019, https://neoteric.eu/blog/10-use-cases-of-ai-in-manufacturing/

[18] **J. Lee et al.,** "Industrial artificial intelligence for Industry 4.0-based manufacturing systems," *Manufacturing Letters*, 2018.

[19] **G. Jujjavarapu, E. Hickok,** and **A Sinha**, "AI and the manufacturing and services Industry in India"

\https://cis-india.org/internet-governance/files/AIManufacturingandServices_Report_02.pdf

[20] **T.M. Knasel**, "Artificial intelligence in manufacturing: Forecasts for the use of artificial intelligence in the USA," Robotics, vol. 2, 1986, pp. 357-362.

[21] "Artificial intelligence: Future & present in manufacturing industry," https://blog.tyronesystems.com/artificial-intelligence-future-and-present-in-manufacturing-industry

[22] "Machine learning & artificial intelligence in advanced manufacturing," https://www.rit.edu/gis/coe/sites/rit.edu.gis.coe/files/docs/news-events/Machine%20Learning%20Blog.pdf

[23] **A. Kusiak,** "Smart manufacturing," International Journal of Production Research, vol. 56, no. 1-2, 2018, pp. 508-517.

[24] **M. N. O. Sadiku**, **O. D. Olaleye**, and **S. M. Musa**, "Smart manufacturing: A primer," International Journal of Trend in Research and Development, vol. 6, no. 6, November-December 2019, pp. 9-12.

[25] **K. S. Prakasha**, **T. Nancharaihb**, and **V.V.S. Rao**, "Additive manufacturing techniques in manufacturing: An overview," Materials Today: Proceedings, vol. 5, 2018, pp. 3873–3882.

[26] **J. P. Davim**, Artificial Intelligence in Manufacturing Research. Nova, 2010.

[27] **R. K. Miller** and **T. C. Walker**, Artificial Intelligence Applications in Manufacturing. Prentice Hall, 2nd edition, 1989.

CHAPTER 10

AI IN AGRICULTURE

"It is difficult to think of a major industry that AI will not transform. This includes healthcare, education, transportation, retail, communications, and agriculture. There are surprisingly clear paths for AI to make a big difference in all of these industries." - *Andrew Ng*

10.1 INTRODUCTION

Agriculture is one of the most fundamental human activities. It is a key economic sector of any nation. It provides food required for human survival. From labor employment to contribution to national income, agriculture contributes immensely to the economy. It is the second largest industry after Defense. Figure 10.1 shows the lifecycle of agriculture.

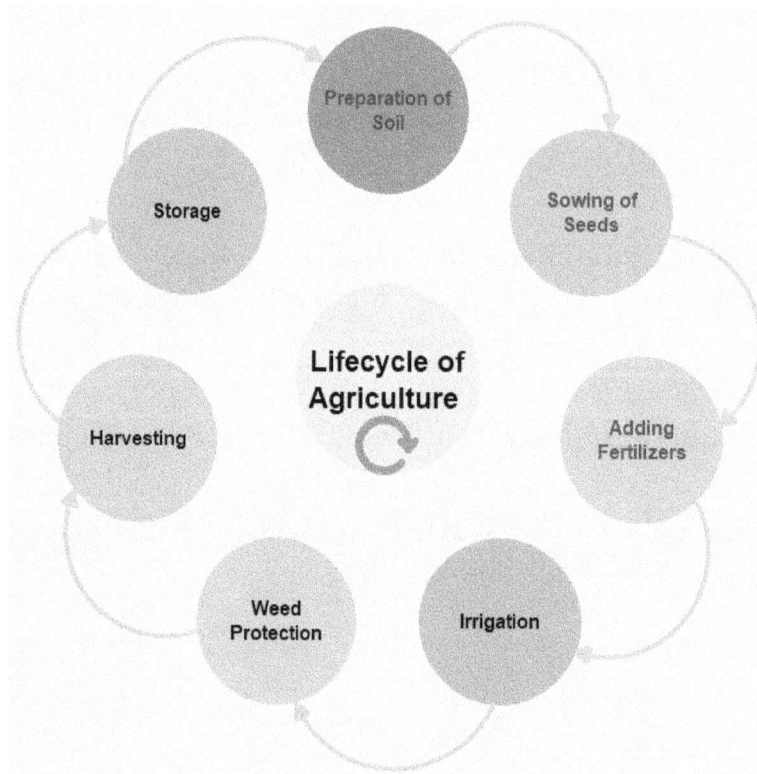

Figure 10.1 Lifecycle of agriculture.

The agricultural industry faces various challenges such as lack of effective irrigation systems, extreme weather conditions, global climatic changes, the rising population, scarcity and increasing labor costs, knowledge gap between farmers and technology, food scarcity, and wastage. The traditional methods used by the farmers are inadequate to solve the difficulties and meet the increasing demand for food by a soaring global population. To meet these challenges, agriculture industry requires investing in new technologies. The agriculture sector is undergoing a transformation driven by new technologies, which have caused a great revolution in agriculture and have made agricultural processes more efficient and precise. Modern agriculture seeks ways to conserve water, use nutrients and energy more efficiently, and adapt to climate change. [11,2].

Digital transformation is disrupting the agricultural world. The application of artificial intelligence (AI) in agriculture has been widely considered as one of the most viable solutions to address food inadequacy. In essence, AI is a technology which functions like a human brain. This technology is based on mimicking how human brain thinks, how humans learn, make decisions, and work while solving a problem. The goals of artificial intelligence include learning, reasoning, and perception.

AI is making a huge impact in all sectors of the economy. It has penetrated business, healthcare, engineering, education, agriculture, industry, security, military, etc. Some of the areas of applications of AI are illustrated in Figure 10.2.

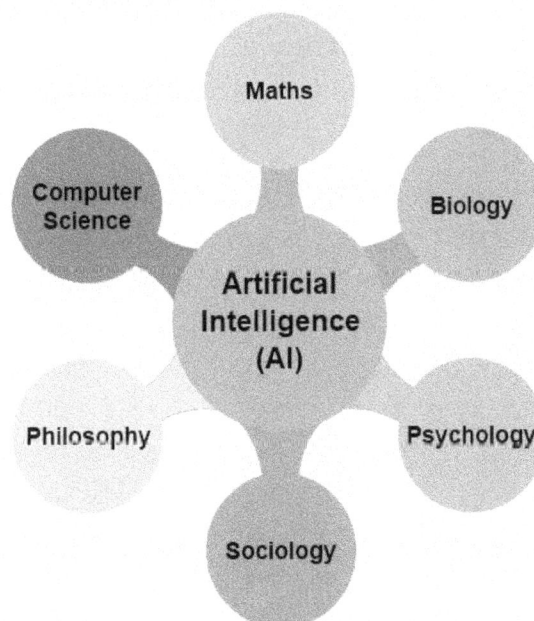

Figure 10.2 Some areas of applications of AI.

Scientists have used AI to develop self-driving cars and chess-playing computers, and now in agriculture. The AI-powered technologies can assist the agriculture sector to yield healthier crops, control pests, and monitor soil. Farmers use AI for in various areas such as precision agriculture, crop monitoring, soil composition, and temperature control in growing areas.

Artificial intelligence is one of the emerging technologies in the field of agriculture which tries to simulate human reasoning in intelligent systems. It is making a revolution in agriculture by replacing inefficient traditional methods with more efficient AI-based methods. AI is used in agriculture in various ways such as automation, robots, drones, soil and crop monitoring, and predictive analytics [3].

This chapter provides various applications of AI tools in agriculture. It begins by giving a brief review of AI. It addresses the roles of AI and machine learning in agriculture. It covers some specific applications of AI in agriculture. It highlights some benefits and challenges of AI in agriculture. It also presents the global application of AI in agriculture and makes some predictions on the future role of AI in agriculture. The last section concludes with comments.

10.2 REVIEW ON ARTIFICIAL INTELLIGENCE

In simple terms, artificial intelligence (AI) refers to computer systems that mimic human cognitive functions. The term "artificial intelligence" (AI) was first used at a Dartmouth College conference in 1956. The main goal of AI is to enable machines to perform complex tasks that typically require human intelligence [4]. AI is now one of the most important global issues of the 21st century. It is poised to disrupt our world and change processes and developments in the field of science, engineering, education, business, and agriculture in recent years.

AI is not a single technology but a range of computational models and algorithms. AI is a collection of techniques that enables computer systems to perform tasks that would otherwise require human intelligence [5].
The major disciplines in AI include:

- Expert systems
- Fuzzy logic
- Neural networks
- Machine learning

- Deep learning
- Natural Language Processors
- Robots

These AI tools are illustrated in Figure 10.3.

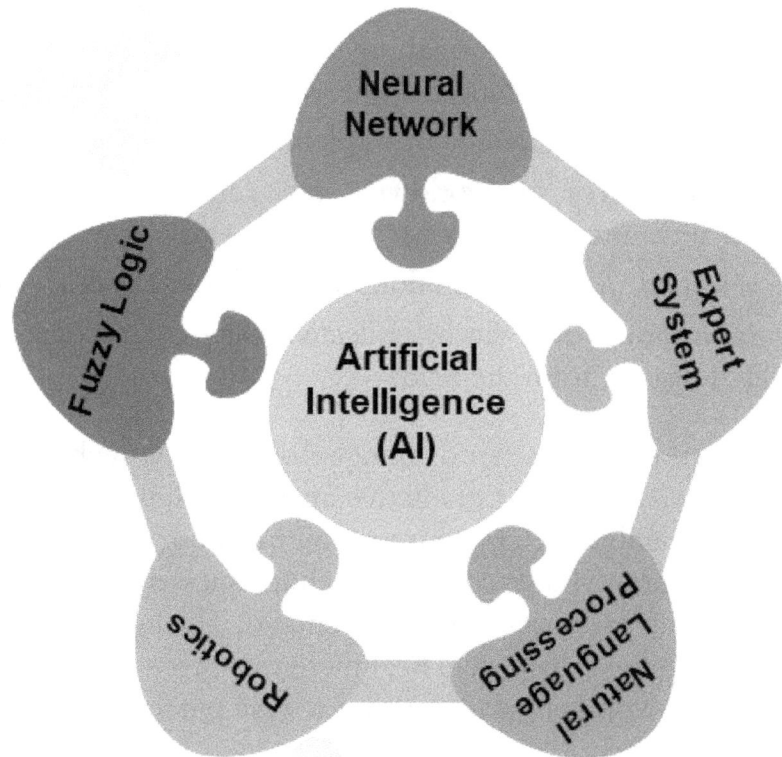

Figure 10. 3 Branches of artificial intelligence.

Each AI tool has its own advantages. Using a combination of these models, rather than a single model, is recommended. AI systems are designed to make decisions using real-time data. They have the ability to learn and adapt as they make decisions. Artificial intelligence is no longer just an academic field; machine learning and deep learning are becoming mainstream technologies that agriculture industry can harness [6]

10.3 AI IN AGRICULTURE

Modern agriculture seeks to reduce cost and minimize environmental impact. Although the integration of AI in agriculture has been relatively recent, its role in agriculture has been much significant than in any other field. AI is transforming agriculture in many ways. Farmers are relying on AI technology in their crop production. Some companies are leveraging computer vision and deep learning algorithms to process the data captured by drones. Food producers are using AI to sort products and reduce labor. The major factors driving the growth of the AI in agriculture market include [7]:

- the growing demand for agricultural production owing to the increasing population

- rising adoption of information management systems and new advanced technologies for improving crop productivity
- increasing crop productivity by implementing deep learning techniques
- growing initiatives by worldwide governments supporting the adoption of modern agricultural techniques
- crop selection: AI-based solutions are ideal for selecting crops
- crop monitoring: data can be collected using technologies like IoT, drones, and satellite imaging, from the fields

The combination of the Internet of things (IoT) and AI technologies, such as machine learning, computer vision, and predictive analytics, allows farmers to analyze real-time data of weather conditions, temperature, soil moisture, plant health, and crop prices in real time. Agricultural AI and cognitive technologies enable farms across the world to run more efficiently and meet our food demand. The agriculture industry is turning to AI-based technologies to help yield healthier crops, control pests, monitor soil, organize data, help with the workload, and improve a wide range of agriculture-related tasks.

10.4 MACHINE LEARNING IN AGRICULTURE

Today, artificial intelligence is narrow and mainly based on machine learning. Machine learning (ML) is a research field in artificial intelligence and statistics that develops computational methods that can be used to learn from data and to predict with new data. It is a type of AI that provides computers with the ability to learn without being explicitly programmed. "Learning" denotes gaining of knowledge and understanding from instruction or experience. Machine learning uses algorithms to parse data, learn from it, and draw conclusions without human intervention. Many industries have realized great benefits from using machine learning to increase reliability, productivity, and the safer operations of their machines. ML algorithms are used to identify the sense of strengths and weaknesses in soil.

In agriculture, we can leverage ML tools that have worked so effectively for other industries. ML has intelligence built in it. It allows for much higher precision, enabling farmers to treat plants and animals almost individually. ML techniques often result from the need for intelligent solutions to practical tasks such as classification, regression or clustering. There are three types of learning/algorithms in ML: supervised learning, unsupervised learning, and reinforcement learning [8,9]. Various ML algorithms can be applied to agriculture. These include [10]:

- **Simple Linear Regression**: This models a linear relationship between one predictor variable known as predictant and one independent variable known as predictor

- **Multiple Linear Regression**: This models a linear relationship between one variable known as predictant with many variables known as predictors.

- **Multilayer perception**: This uses supervised learning technique known as backward propagation for training an artificial neural network (ANN).

- **Support Vector Machine**: This is a supervised ML algorithm that is used for classification and regression.

ML techniques may result in data insights that increase production efficiency. The applicability of ML in agriculture has a lot of benefits, from tilling soil, plant breeding, spraying fertilizer, crop disease detection and to water usage for irrigation, and harvesting [11]. Some applications of ML in the agri-food space are illustrated in Figure 10.4.

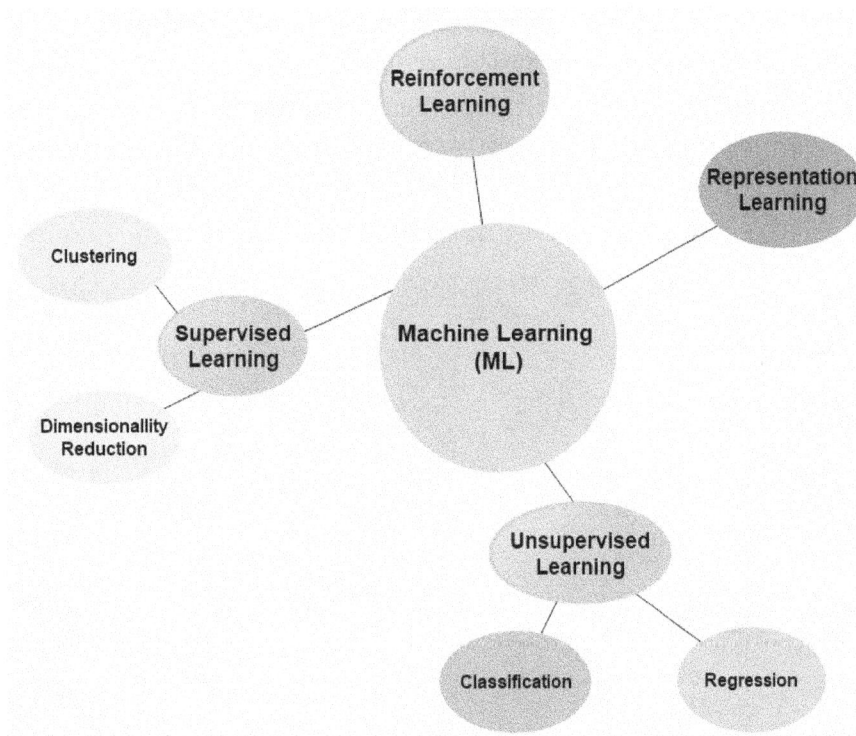

Figure 10.4 Some applications of machine learning in agriculture.

10.5 APPLICATIONS OF AI IN AGRICULTURE

AI tools have taken modern agriculture to a different level. They have enhanced crop production and improved real-time monitoring, harvesting, processing, and marketing. Some representative examples of popular applications of AI in agriculture and farming include the following [12,13].

Robots: A robot is a smart system that contains sensors, control systems, power supplies, and software all working together to perform a task. Robots are commonly used in dangerous environments where humans cannot survive such as defusing bombs, finding survivors in unstable ruins, welding, and exploring mines and shipwrecks. Today, robots perform important functions in homes, hospitals, industries, military institutions, education, entertainment, and outer space. Automated systems using agricultural robots and drones have made a tremendous contribution in agriculture. Farmers are using robots that can easily perform multiple functions. Some robots are designed to weed, hoe, spray, keep vigilance, and assist during harvesting. Agricultural bots are specifically designed robots to implement automation in agriculture. Typically, a robot is trained to control weeds and harvest crops at a faster pace than humans. The flying robot is often powered by computer vision. Although robots will be used in agriculture, the human worker will still need to supervise and do other tasks for which the robots are not trained. Figure 10.5 shows how computer vision teaches the robots to pick fruit [14].

Figure 10.5 Computer vision teaches the robots to pick fruit [14].

Drones: DRONE (Dynamic Remotely Operated Navigation Equipment) is commonly referred as unmanned aerial vehicle (UAV). A drone is used in farming to help increase crop production and monitor crop growth. It is a pilotless aircraft, designed to collect more accurate information than airplanes or satellites. Once the drone captures and processes the data, the data is sent to farmers in a readable format for management decisions. Drones are based on the innovations of sensors and microcontrollers. They are used for spraying and crop-monitoring. They are effective in the situations of cloudy climate and inaccessibility to a field of tall crops. They are capable of capturing far more land in much less time than humans. Using drones, AI-powered cameras can take images of the entire farm. Agriculture drones are typically used for aerial photography in livestock operations, spraying, drought assessment, monitoring, etc. Drones are being implemented in agriculture for crop health monitoring, irrigation equipment monitoring, weed identification, herd and wildlife monitoring, and disaster management. Drone data is a powerful tool to help farmers visualize your fields. The drone captures data from fields and transfers the data to a computer so that experts can analyze it. Drones can analyze and collect data at a much faster rate than humans. Drone technology is giving agriculture a high-tech makeover. Figure 10.6 illustrates how drones are used in agriculture to gather information [15].

Figure 10.6   Drones are used in agriculture to gather information [15].

Chatbots: Chatbots (also known as a talkbots or chatterbots), as part of AI devices, are computer programs designed to carry on a dialogue with users using natural languages. They simulate human conversation via text or speech. They are also known as conversational agents, interactive agents, virtual agents, virtual humans, or virtual assistants. Specialized chatbots are becoming more prevalent in our daily lives, especially in agriculture. They are basically conversational virtual assistants who automate interactions with end users. AI-powered chatbots can help users understand natural language and interact with users in a more personalized way.

Precision Farming: AI systems are helping to improve the overall harvest quality and accuracy. Precision agriculture (or precision farming) is about providing more accurate farming techniques for planting and growing crops. It is the use of information and communication technology (ICT) along with best agricultural practices. It may boost agricultural production while reducing harmful impact on the environment. It can make a big difference to food production facing the challenge of a rising world population. AI applications in agriculture have developed applications and tools which help farmers by providing them proper guidance on water management, crop rotation, timely harvesting, type of crop to be grown, etc. [16]. Precision farming uses AI technology to aid in detecting diseases in plants, pests and poor plant nutrition on farms. Precision farming will optimize the processes of monitoring the state of soil and crops. AI-enabled technologies can predict weather conditions, temperature, precipitation, and wind speed. Microsoft precision agriculture attempts to democratize AI for farmers around the world.

Monitoring: This allows farmers to monitor soil and crop conditions. Farmers leverage sensors and various IoT-based technologies to monitor crop and soil health. Satellites can be used for monitoring crop health and sustainability. AI also allows farmers to monitor the health of animals and manage the number of livestock. Smart devices can help control their movement and track feed requirements.

Weather Forecast: Agriculture across the world is dependent on climate. Weather tracking and forecasting are important applications of AI in agriculture. They involve collecting up-to-date information of prevailing weather conditions such as temperature, rain, wind speed and direction, and solar radiation. Various devices such as handheld instruments, sensors, GPS and on-field weather stations are used for tracking weather. The availability of real-time information helps farmers in making timely decisions.

Predictive Analytics: This is one of the most promising areas for the farming industry. AI has an array of tools to predict changes in weather patterns, soil conditions, pest infestation or soil quality, and composition in order to improve planning and farm management. Predictive analysis is used in a variety of ways. These include solutions that predict crop yields using ML algorithms and AI algorithms can evaluate the creditworthiness of farm borrowers. The crop yield prediction is beneficial for marketing strategies and crop cost estimation. Advances in AI-based data analytics help farmers protect natural resources like land, air, and water, and reduce the amount of inputs needed for successful harvests.

Other areas of applications of AI in agriculture include indoor farming, weed management, disease management. detection of pests, intelligent irrigation, labor efficiency, crop diversity, supply chain management, risk management, pest control, gardening sector, and food manufacturing.

10.6 BENEFITS

The main benefits of AI in agriculture are its flexibility, high performance, accuracy, and cost-effectiveness. AI-based technologies help to improve efficiency in all the areas, while the conventional farming machineries lack in efficiency. With AI-based solutions, farmers can meet the world's need for increased food without depleting natural resources. AI technology will improve the way we think, the way we explore new horizons, whether space or the ocean. AI and ML are gradually replacing outdated ways of planning and forecasting.

Some advantages of implementing AI in agriculture include [17]:

• AI provides more efficient ways to produce, harvest, and sell essential crops.

• AI can help farmers get more from the land while using resources more sustainably.

• AI is an excellent choice for automated irrigation and weeding. Manual irrigation is based on soil water content. AI takes into account humidity, wind speed, solar radiation, stage of crop, plant density, etc. and generates the most optimal irrigation schedule.

• AI can help humanity solve one of its biggest challenges: feeding an additional 2 billion people by 2050.

• AI technology can help farmers overcome many issues like climate variations, water availability, pest control, etc.

• AI implementation emphasis on checking defective crops and improving the potential for healthy crop production.

• The growth in artificial intelligence technology has strengthened agro-based businesses to run more efficiently.

- AI is being used in applications such as automated machine adjustments for weather forecasting and disease or pest identification.

- AI can improve crop management practices and help many tech businesses invest in algorithms that are becoming useful in agriculture.

- AI solutions have the potential to solve the challenges farmers face such as climate variation, an infestation of pests and weeds that reduces yields.

- AI in agriculture optimizes the agriculture industry by decreasing workloads, analyzing harvesting data and improving accuracy.

- AI solves labor challenges by augmenting or removing work and reducing the need for large numbers of farm workers.

- AI is emerging as a significant solution towards the improved agricultural productivity. Productivity gains can improve global food supply and lead to indefatigability, minimization of errors, and consistency of work quality.

- AI in farming has empowered the farmers to be future-ready.

10.7 CHALLENGES

While AI has brought a lot of promise, it has also brought some challenges. In spite of the fact that agriculture is a data-centric industry, there are significant challenges on the way to the global adoption of AI. Agriculture is one of the most uncontained environments to manage. The agricultural industry faces many tremendous challenges such as lack of effective irrigation systems, issues with plant monitoring due to crop height, extreme weather conditions (global warming), worry due to price fluctuation of the crop, omnipresent uncertainty, and the need to produce more food using fewer resources. These challenges are driving farmers and agro companies to find ways that are more efficient than the traditional methods. As a result, AI is emerging and promising to help farmers solve these challenges and get more from the land while using resources more sustainably.

AI is far from a silver bullet. Securing access to AI on a global level may pose some challenges. There is a gap between farmers and AI engineers. The cost of technology such as drones has made it unavailable to an average farmer. AI solutions in agriculture will require new, common terminologies to be agreed upon globally. AI will replace jobs in factories and in agriculture such as hand harvesters.

Other challenges of the application of AI-based techniques in agriculture include [18]:

- Possible uneven future distribution of mechanization
- Discrepancies between control experiments and actual implementation
- Security and privacy
- AI-enabled technologies require massive investment to buy them.

10.8 GLOBAL AI IN AGRICULTURE

Agriculture plays a vital role in the global economy as well as the economy of each nation. Worldwide, agriculture is a $5 trillion industry. Most nations are committed to improving the profitability and efficiency of the agricultural sector. AI- based technologies will assist farmers to obtain more from their limited land and other resources. Global AI in agricultural market is segregated into predictive analysis, computer vision, and machine learning. Here we consider how AI is used in agriculture worldwide.

United States: North America has been witnessed as the largest market for AI in agriculture. The presence of many agriculture technology providers such as IBM, Microsoft, John Deere, Granular, Autonomous Tractor Corporation, Fendt, and Blue River Technology in US is driving the growth of the North American AI in agriculture market. These companies are pushing the boundaries of technology when it comes to AI's role in agriculture.

A United Nations agency says the United States, China, and Japan are leading the world in developing AI technology. The study is based on international patent requests, legal findings, scientific publications, and other related information [19].

India: The agriculture and allied industries are the backbone of the India's economy. AI-based technologies are gradually entering Indian agriculture and have started to shape agricultural practices in India. Being the fastest growing economy with the second largest population in the world, India has a significant stake in the AI revolution. India's government has identified agriculture as one of the key areas where AI can enable development and greater inclusion. The government has also set a target of doubling of farmer's income by the year 2022. Indian farmers are now adopting smart farming strategies powered by AI enabled sophisticated technologies, autonomous tractors fitted with GPS and various other sensors including digital cameras in more efficient ways than ever before. AI-enabled solutions in agriculture help farmers improve crop productivity, monitoring soil health, optimize pest and weed management, manage irrigation. AI algorithms are being used to monitor crop and soil health [20]. The most popular applications of AI in Indian agriculture appear to fall into three major categories [21]:

(1) Crop and Soil Monitoring – Companies are leveraging sensors and various IoT-based technologies to monitor crop and soil health; (2) Predictive Agricultural Analytics – Various AI and machine learning tools are being used to predict the optimal time to sow seeds, get alerts on risks from pest attacks, and more; (3) Supply Chain Efficiencies– Companies are using real-time data analytics on data-streams coming from multiple sources to build an efficient and smart supply chain.

China: About half of China's farms are small-scale operations. China is growing at a swift pace when it comes to AI. The rising adoption of computer vision technologies and deep learning for AI applications is one of the key drivers for the market expansion in the APAC countries such as China.

Norway: Norway is a highly digital country. It has a well-developed regulatory framework for digital technologies and services. But they are often using technologies developed in other countries. The Norwegian Government recently launched a National Strategy for Artificial Intelligence that sets out how the country will develop AI-based technology that will promote sustainable development. Among other things, Norway is using artificial intelligence to make it easier to maintain the electricity grid [22].

New Zealand: Agriculture and horticulture play a major role in New Zealand's export economy. However, the agriculture industry faces significant ongoing challenges including climate change, low productivity growth, labor shortages, increasing regulation, and environmental sustainability. AI in agriculture is in its early days and in an embryonic stage in New Zealand. The rapid development of AI technologies presents major opportunities and challenges for this country: from creating world leading AI businesses, nurturing a pool of talented AI engineers, and applying AI technologies to agriculture. The AI Forum of New Zealand is a non-government association with a mission to harness the potential of AI to help bring about a prosperous and inclusive future for the country. The Forum brings together citizens, business, academia, and the government to connect, promote, and advance the AI technology to ensure a prosperous future [23].

10.9 FUTURE OF AI IN AGRICULTURE

Artificial intelligence has started to live up to its promise of delivering real value, driven by recent advances in the availability of relevant data, computation, and algorithms. The reality of today which was a yesteryear dream shows how rapidly AI technology is bringing in the revolution in farming. Artificial intelligence is a solution for the future.

AI can drive agricultural revolution at a time when the world must produce more food using fewer natural resources. As AI permits the sphere of agriculture. The future promises more food with less consumption of natural resources [24]. Here are three exciting trends in agriculture to watch out for the coming years.

Automation and Robotics: Automation is emerging in an effort to help address challenges in the labor force. Agriculture automation is the main concern and emerging issue for every nation. Traditional farming is not sufficient to serve the increasing global demand for food. Automation of farming practices has increased the gain from the soil and increase productivity. Thus, automation in the agriculture sector is necessary and there are many ways to implement it in practice. Different automation practices like AI, ML, deep learning, robots, and IoT have started penetrating in this sector. It is needless to say that these methods have their own advantages and disadvantages [25]. As AI is the mainstay of robotics, increasing adoption of robotics in agricultural sector is expected to drive the market. The gradual shift of farmers towards robots, automated machines, drones, chatbots or agrobots, and growing trends of precision farming will spur the market. The self-driving tractors can perform multiple activities without any human. Agriculture industry will soon to be hiring more robot employees.

Indoor Farming: Indoor farming powered by AI may be the key in fighting climate change. It is the future of farming. It is attracting a whole new breed of farmers. Scientists at Microsoft are revolutionizing conventional farming by sowing, growing, and harvesting seeds under controlled temperatures powered by AI. With the help of AI, vegetables, fruits, flowers, mushrooms, herbs, etc. can be grown under the favorable combination of blue, red, and white light. Vertical cropping can reduce water usage, make efficient land usage, and can be cultivated in building within urban areas.

Universal Access: The future of AI in agriculture will need a major focus on universal access. Small farmers around the world follow traditional farming practices due to lack of access to scientific understanding of crop lifecycle, pests, quality metrics, the latest micro-fertilizers, and related emerging technologies. Despite the promise of AI tools (such robots, drones, and chatbots), many farmers cannot afford them. In the coming years, we will see more discoveries made in the field of agricultural AI. In order for these farmers to use the right tools, the discoveries should be readily available and affordable to all farmers.

Other trends involve precision farming/agriculture, agriculture monitoring, and predictive analysis.

10.10 CONCLUSION

With the widespread use of artificial intelligence across all industries, it is not surprising that AI is gradually transforming agriculture. Artificial Intelligence has brought an agriculture revolution. It is already at work in agriculture industry and is here to stay. Through AI technology, farmers are well equipped to tackle the huge challenges of a growing global population and world hunger. AI will be a powerful tool that can help organizations cope with the increasing amount of complexity in modern agriculture.

Through robots, monitoring of soil, and by making data supported predictions, AI is emerging in the field of agriculture. Since enough food will be produced locally, nations will not need to import food products internationally. Data-driven approaches integrating AI and ML with big data technologies and high-performance computing could drive agricultural productivity while minimizing its environmental impact. We anticipate that the agricultural industry will continue to see steady adoption of AI and will continue to monitor this trend. For more information about AI in agriculture, one should consult the book in [26] and the following related journal: Artificial Intelligence in Agriculture.

REFERENCES

[1] **M. N. O. Sadiku, S. M. Musa**, and **A. Ajayi-Majebi**, "Artificial intelligence in agriculture," International Journal of Trend in Scientific Research and Development, vol. 5, no. 2, Jan.-Feb. 2021, pp. 721-725.

[2] **N. C. Eli-Chukwu**, "Applications of artificial intelligence in agriculture: A review," Engineering, Technology & Applied Science Research, vol. 9, no. 4, 2019, pp. 4377-4383.

[3] **Y. Wang**, "Interdisciplinary study for future artificial intelligence development," February 2018, https://medium.com/anth374s18/interdisciplinary-study-for-future-artificial-intelligence-development-28277e6ce9a5

[4] **H. Chen, L. Li**, and **Y. Chen**, "Explore success factors that impact artificial intelligence adoption on telecom industry in China," Journal of Management Analytics, 2020.

[5] **S. Greengard**, "What is artificial intelligence?" May 2019, https://www.datamation.com/artificial-intelligence/what-is-artificial-intelligence.html

[6] **"Artificial intelligence in agriculture:** Using modern day ai to solve traditional farming problems," November 2020, https://www.analyticsvidhya.com/blog/2020/11/artificial-intelligence-in-agriculture-using-modern-day-ai-to-solve-traditional-farming-problems/

[7] "Rapid adoption of artificial intelligence in agriculture," August 2019, https://www.futurefarming.com/Smart-farmers/Articles/2019/8/Rapid-adoption-of-artificial-intelligence-in-agriculture-461266E/

[8] **K. Kaur**, "Machine learning: Applications in Indian agriculture," International Journal of Advanced Research in Computer and Communication Engineering, vol. 5, no. 4, April 2016, pp. 342-344.

[9] **M. N. O. Sadiku, S. M. Musa**, and **O. S. Musa**, "Machine learning," International Research Journal of Advanced Engineering and Science, vol. 2, no. 4, 2017, pp. 79-81.

[10] **H. D. Aparna, K. S. Kavitha**, and **C. Kavitha**, "Use of machine learning techniques to help in predicting fertilizer usage in agriculture production," International Research Journal of Engineering and Technology, vol. 4, no. 5, May 2017, pp. 1868-1869.

[11] **M. N. O. Sadiku, C. M. M. Kotteti**, and **S. M. Musa**," Machine learning in agriculture," International Journals of Advanced Research in Computer Science and Software Engineering, vol. 8, no. 6, June 2018, pp. 26-28.

[12] **N. N. Misra et al.**, "IoT, big data and artificial intelligence in agriculture and food industry," IEEE Internet of Things Journal, 2020.

[13] **T. Talaviya et al.**, "Implementation of artificial intelligence in agriculture for optimisation of irrigation and application of pesticides and herbicides," Artificial Intelligence in Agriculture, vol. 4, 2020, pp. 58-73.

[14] **V. Kumar**, "The power of artificial intelligence in agriculture," July 2019, https://www.analyticsinsight.net/the-power-of-artificial-intelligence-in-agriculture/

[15] **A. Garg**, "Applications of artificial intelligence in agriculture and farming," May 2020.

[16] **M. N. O. Sadiku, Y. Wang, S. Cui, S. M. Musa**, "Precision agriculture: An introduction," International Journal of Advanced Engineering and Technology, vol. 2, no. 2, May 2018, pp. 31-32.

[17] **J. Gupta**, "The role of artificial intelligence in agriculture sector," October 2019, https://customerthink.com/the-role-of-artificial-intelligence-in-agriculture-sector/

[18] **J. Zha**, "Artificial intelligence in agriculture," Journal of Physics: Conference Series, 2020.

[19] **"UN study: China, US, Japan lead world AI development,"** https://learningenglish.voanews.com/a/un-study-china-us-japan-lead-world-ai-development/4769094.html

[20] "Artificial intelligence (AI) - A revolution in India's agriculture," November 2020, https://www.ibef.org/research/newstrends/artificial-intelligence-ai-a-revolution-in-india-s-agriculture#:~:text=AI%2Denabled%20solutions%20in%20agriculture,best%20price%20for%20the%20farmer.

[21] "Artificial intelligence in Indian agriculture – An industry and startup overview," https://emerj.com/ai-sector-overviews/artificial-intelligence-in-indian-agriculture-an-industry-and-startup-overview/

[22] **J. Sortino**, "This is how Norway puts artificial intelligence to use," February 2020, https://www.theexplorer.no/stories/technology/this-is-how-norway-puts-artificial-intelligence-to-use/?gclid=EAIaIQobChMIudOKtdXn7wIVijizAB0FVwccEAAYAyAAEgJLP_D_BwE

[23] **A. A. Iahiko**, "Artificial intelligence for agriculture in New Zealand," https://aiforum.org.nz/wp-content/uploads/2019/10/Artifical-Intelligence-For-Agriculture-in-New-Zealand.pdf

[24] "The future of artificial intelligence and agriculture," November 2018, https://medium.com/@ODSC/the-future-of-artificial-intelligence-and-agriculture-540c39208df6

[25] **K. Jha et al.**," A comprehensive review on automation in agriculture using artificial intelligence," Artificial Intelligence in Agriculture, vol. 2, June 2019, pp. 1-12.

[26] **R. Singh et al.**, Artificial Intelligence in Agriculture. New India: New India Publishing Agency, 2021.

CHAPTER 11
AI IN FOOD INDUSTRY

"For the first 50 years of your life the food industry is trying to make you fat. Then, the second 50 years, the pharmaceutical industry is treating you for everything."
-Pierre Dukan

11.1 INTRODUCTION

Food is an essential part of our lives. The food industry is the basic and important to every nation. It is one of the seventeen national critical sectors of US economy. It plays a crucial role in public health, food safety, food security, social development, and nutrition. It covers diverse activities including food supply, production, harvesting, processing, packaging, transportation, distribution, consumption, and disposal. The food industry in its entirety is not one industry but a collection of several types of industries producing a diverse range of food products. The food industry is dominated by multinational corporations such as Krafts Foods, Cadbury, Heinz, Nestlé, Food World, DuPont, McDonalds, Pizza Hut, and KFC [1].

Agriculture, forestry, and fisheries can provide nutritious food for the world. The global food industry continues to grow steadily. (Coronavirus is impacting the global food industry.) The global population is increasing and is widening the gap between food supply and demand. Today, the global population exceeds seven billion. The rising global demand for food can only be met using technology. It also requires developing strong business relationships with farmers and agronomic experts across the globe.

The food industry has always been at the forefront of adopting emerging technologies to improve the sector. Artificial intelligence (AI), with the capacity to make computers to learn from experience, is playing a predominant role in the food industry. Artificial Intelligence (AI) is poised to revolutionize the food industry. AI is one of emerging technologies that can literally transform the agricultural landscape in the years to come and take agriculture to new heights [2]. AI technology can be implemented in all stages of the food supply chain, resulting in an overall improvement and significant increase in efficiency. Forecasting global demands and delivering safe food products can be done using machine learning and deep learning, two of the most used algorithms in the field of AI.

This chapter provides an introduction on the applications of AI in food and beverage industry. It begins by giving a brief review of AI. It covers some research projects on the use of AI in food industry. It mentions some specific applications of AI in the food industry. It highlights some benefits and challenges of AI in the industry. It presents the global application of AI in the food industry. It makes some predictions on the future role of AI in food and beverage industry. The last section concludes with comments.

11.2 REVIEW ON ARTIFICIAL INTELLIGENCE

Artificial intelligence (AI) refers to computer systems that mimic human cognitive functions. It is a field of computer science that deals with intelligent machines. The term "artificial intelligence" was first used at a Dartmouth College conference in 1956. The main goal of AI is to enable machines to perform complex tasks that typically require human intelligence [3]. In simple terms, AI attempts to clone human behavior. An important feature of AI technology is that is can be added to existing technologies. AI is now one of the most important global issues of the 21st century. It is poised to disrupt our world and change processes and developments in the field of science, engineering, education, business, and agriculture.

AI is not a single technology but a range of computational models and algorithms. AI is a collection of techniques that enables computer systems to perform tasks that would otherwise require human intelligence [4]. The major disciplines in AI include:

- Expert systems
- Fuzzy logic
- Neural networks
- Machine learning
- Deep learning
- Natural Language Processors
- Robots

These AI tools are illustrated in Figure 11.1.

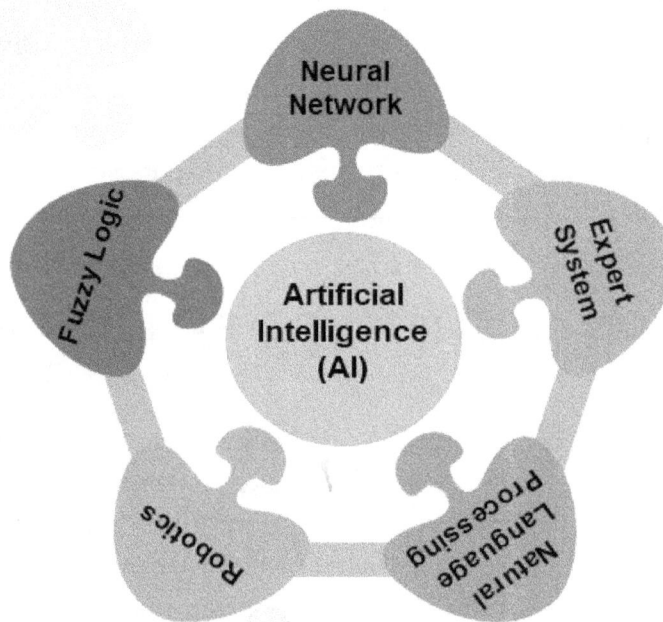

Figure 11.1 Branches of artificial intelligence.

Each AI tool has its own advantages. Using a combination of these models, rather than a single model, is recommended. AI systems are designed to make decisions using real-time data. AI algorithms, such as machine learning and deep learning, are most commonly used to make intelligent predictions. They have the ability to learn and adapt as they make decisions. AI is no longer just an academic field; machine learning and deep learning are becoming mainstream technologies that manufacturing companies can harness. Figure 11.2 illustrates the six Ws of artificial intelligence.

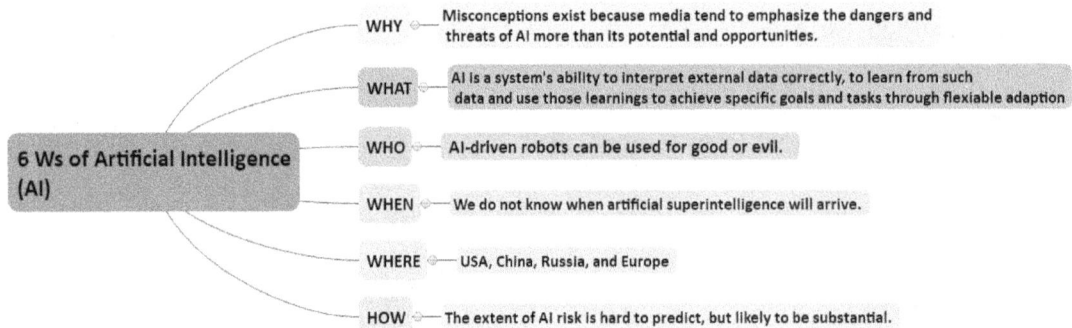

Figure 11.2   The six Ws of artificial intelligence.

AI systems are designed to make decisions using real-time data. They have the ability to learn and adapt as they make decisions [5,6].

11.3 AI RESEARCH PROJECTS IN FOOD INDUSTRY

The application of AI in the food and beverage industry is already transforming the way we think about food production, food manufacturing, food processing, food quality, food delivery, food consumption, and food storage [7]. Several food companies have incorporated machine learning, deep learning into food and beverage products and services. Here are the latest AI research projects related to the food industry [8]:

- Predicting food security outcomes
- Food identification
- Ranking food preferences
- Automatic surface area and volume prediction of food
- Generating images of food based on recipes text
- Automatically assign a collective restaurant star rating based on customer reviews of food
- Food recommender system based on the user's history, ingredients, and image of a recipe
- Visual identification of fraudulent foodstuff products
- Food recognition using partially labeled data
- Recipe generation from food images
- Recognizing eating gestures by tracking wrist motion
- Real-time detection of foodborne illness
- Automated food label quality assessments
- Plant seedlings classification
- Automatically assessing the health of animals, fish, etc.
- Identification of leaf diseases

11.4 APPLICATIONS OF AI IN FOOD INDUSTRY

In addition to the research projects mentioned above, key areas application of AI in food industry include production, product development, product customization, marketing, manufacturing, robotics, and processing of food products. AI can also be applied in restaurants, bar, and cafe businesses [9-12].

Agriculture: Agriculture is facing major challenges and AI is regarded more and more as a solution to these challenges. There is an unquestionable growing tendency in the adoption of AI in agriculture. Making the right decisions in agricultural activities is less than half the way needed to achieve desired results. The intelligent agents implementing measures into practice are seen more often as a feasible solution for the future. Agriculture automation is the main concern of every country. It has benefited from different automation practices like IoT, wireless communications, artificial intelligence, machine learning, and deep learning. However, the use of technologies in agriculture is still in infancy [13,14].

Food Processing: Food processing is a complicated business and labor intensive because it involves sorting the food or raw materials coming from the farm. A major challenge in food processing plants is that the feedstock is often not uniform. Every orange, carrot, tomato, potato, apple, etc. is slightly different.: Food processing is influenced by various factors such as quality control, type of food, current trends, customer psychology, and human wellness. To handle the market's demand and to yield rapid production, the food industry adopted few food processing techniques. Food processing in many places is not fully automated. The food processing industry is benefiting from AI, which is helping everything from sorting foods, maintaining health and safety compliances, developing new products, and improving the supply chain. The technology is basically helping to streamline work processes, making the work of employees easier and making operations more efficient. Food processing industry can guarantee complete hygiene and high food quality by automating their processes as much as possible. For example, a potato processing company has used AI to determine which potatoes will produce the least waste when cut into French fries and which ones would work best for potato crisps [15]. Figure 11.3 illustrates food processing [16].

Figure 11.3   Food processing [16]

Food Manufacturing: AI can be used by food companies in food manufacturing. Manufacturing of large productions of goods requires large, complicated, and intricately constructed mechanisms. AI in food manufacturing helps to monitor every stage of this process and makes predictions about cost and stock management. Machine learning allows identifying factors affecting the quality and causing flows in the manufacturing process and monitors the path of products from where they are produced and, ultimately, to where customers get it, ensuring transparency.

Food Production: AI is used in advanced applications to make food production more efficient, safe, and profitable. It has enormous potential to optimize production and uncover manufacturing facilities' best operating points. It can enable faster production changeovers and identifying production bottlenecks before they become a problem. AI can optimize your facility's productivity, efficiency, and output.

Food Packaging: AI-powered robotic equipment can accurately perform complex human tasks like the packaging. Robots are meeting the packing and picking demands accelerated by consumers' requests. The complex and labor-intensive nature of the process offers unique potential for intelligent automation.

Food Supply Chain: The food industry needs to supply markets with high-quality food products at reasonable cost to satisfy consumer demands. Supply chains in the food industry are expanding steadily. To enhance food safety, companies can use AI's potential across their supply chain to minimize delays and maximize profit margins by providing close monitoring of every supply chain operation. The technology also enables companies to test and monitor food safety products at every step of the supply chain. AI can track products from the farm to consumers to ensure transparency. AI-powered supply chain management systems can be used to monitor and control the activities in the entire supply chain. AI can be used to minimize delays and maximize profit margins by providing close monitoring of every supply chain operation.

Food Delivery: This is a growing segment of the food industry. The use of AI is a common enabler for most food delivery organizations. Leading food delivery services are using AI to enhance their marketing efforts through automation. AI applications also improve operational efficiency by automating activities like food ordering, dispatching, and billing processes. Food delivery service companies also use AI to enhance their customer services. AI is used to increase efficiency and lower costs during food delivery, working with restaurants to ensure fast delivery of their food.

Restaurants: AI is also being used by many apps to recommend the best restaurants based on the location, taste, and previous choices of the user. Restaurants can stay on top of everything by proper implementation of AI technology into their business. An AI in food service solution offers to unite the data from various food delivery programs. Self-serving systems are being massively taken up by restaurants as long as they enable customers to control the ordering process. In order to reduce the waiting time for the customers and improving their experience, restaurants are now integrating AI-driven kiosks. Some restaurants now make use of chatbots to be able to respond to customer queries quickly and to process their orders [17].

Robotics: Robots are widely used in the food industry, but they are only for privileged big food businesses, still unavailable for small-to-medium businesses. Robots are being used from seeding, spraying water, and harvesting to cutting, processing and packaging of food products [18]. These are drones to deliver orders or robotic hands that can manage many processes in food manufacturing and even cooking. For example, 7-Eleven claims they use drones in their delivery service. Although robotics is still a quite subtle thing to introduce in some factories, it will bring an obvious benefit in the long run. Robots are used for crop cultivation, crop imaging, and in milking cows. The combination of computer vision with robotics provides new innovative approach for farming. Today, robot chefs are food processors and appliances that require minimal supervision in the entire cooking process. As another example, packaging robots can open, fill, pack, seal, and correctly label the package to be sent to the end-user [19]. Figure 11.4 shows the use of robots in food industry [20].

Figure 11.4 Use of robots in food industry [20].

Other applications include food safety, food retailers, food market analysis, optical food sorting, predictive maintenance, tracking and traceability, personalized customer service, and heightened consumer engagement.

11.5 BENEFITS

The implementation of AI in the food and beverage industry is enabling less human errors, less wastage of abundant products, save costs, happier customers, optimize and automate processes, and more personalized orders. AI could utilize vast datasets of detailed agricultural information for the improvement of our food crops, faster than ever before. Adoption of AI can position companies as digital leaders. Competitive pressure is the key driver for food companies to adopt AI. Other benefits include [21,22]:

• **Growing Better Food:** AI could help farmers actually grow better food by creating optimal growing conditions. Businesses can implement intelligent algorithms to improve their quality of food and services which in turn will mean many healthy meals for the customers.

• **Food Safety:** Safety is a major concern in the food processing industry. Reducing the presence of pathogens and detection toxins in food production is a key benefit for employing AI. To ensure safety, robots in food production are sterile. AI solutions display great potential for the optimization of the hygiene and cleaning tasks, which is the factor number one to influence food safety.

• **Food Waste:** AI can minimize food waste. Smart logistics can significantly improve the food supply chain by minimizing food waste.

• **Automated Sorting**: This reduces labor costs, increases the speed of the process, and improves the quality of yields. Proper sorting of food products is one of the major benefits in AI in food industry.

• **Productivity**: Productivity has increased in the food preparation, handling and production in the food industry due to the use of AI technology. AI systems deliver more accurate production lines results with greater speed and more consistency than human workers.

• **Ensuring Personal Hygiene:** Product quality, health, and sanitation issues are major concerns in the food industry. Since cleanliness is a priority, AI makes sure to maximize food processing plant hygiene and ensure that the plant is running according to US government guidelines. Clean-in-place (CIP) systems are programmed to clean equipment in timed cycles. This limits human intervention and the chances of cross-contamination from foodborne pathogens.

- **Achieving Zero Hunger:** AI in the food industry help us respond intelligently to the global demand for foodstuff products. We can address world hunger using AI and machine learning.

11.6 CHALLENGES

While there are many benefits of using AI in food industry, there are some challenges. The challenges that come along with using AI in the food industry include [23,24]:

- **Cost**: Potential adapters of AI in the food industry face numerous difficulties with the price being the biggest among them. The high cost of large-scale deployment in the sector is restricting the market to grow.

- **Integration:** It is not easy to integrate new technologies. This applies for the integration of AI in food companies.

- **Proprietary Data:** With the right proprietary data, food and beverage companies may not be able to build artificial learning models that can perform.

- **Common Fear:** AI will never replace humans in the food industry, as humans will always be needed to oversee operations, repair, and maintain old equipment. The technology can essentially work side-by-side with humans to increase operational efficiency.

- **Shortage of AI Workers:** AI is yet to be ubiquitous due to cost constraints and lack of skilled experts.

11.7 GLOBAL AI IN FOOD INDUSTGRY

The global food industry continues to grow steadily, driven mainly by population growth, rising income levels, and changing lifestyles. The demand in growth for food supplies invokes equivalent production values and sustainable methods. With growing population in the rise, the Food and Agricultural Organization (FAO) of United Nations stated that this population can reach around 9.1 billion by 2050. According to FAO, 793 million people (one in every nine persons) in the world lacks daily food [25]. Besides US and Europe, the adoption of AI is still in the early stages in many nations. Other nations, including Canada, the United Kingdom (UK), China, Estonia, and Singapore are already investing strategically and making rapid advances. Countries such as India and China are making government policies to adopt AI technology to improve the productivity of food supply. Tech giants such as Microsoft and Google are using their technology to support these countries. We now consider how AI is used in the food industry worldwide.

- **United States**: Food processing is one of the major manufacturing industries in the US. Food processing companies are turning to AI technology in order to improve numerous aspects of the process. It is evident that an emerging duopoly with the United States and Chinese companies dominate AI investment. Dallas-based technology company Symphony Retail AI uses AI in the food supply chain to boost productivity, and greatly improve the accuracy of information for better decisions. Tastewise is US startup using AI-based intelligence for food retail.

- **China**: China attracts the major share of global investments in AI and boasts of a booming food delivery market that is ten times bigger that its US counterpart. The Chinese government has developed a national strategy for AI commercial and military applications and aims to make China the dominant global player with an AI sector. For personal hygiene, a Chinese company KanKan created an algorithm-based machine learning system that uses cameras as well as object and facial recognition features to monitor which workers are wearing safety masks. A KanKan subsidiary consisting of AI-enabled cameras in Shanghai's municipal checks that workers are complying with the safety regulations. More recently the company added improved facial recognition abilities to account for the mandatory use of a mask, and new body temperature detection, in line with effects of Covid [26].

- **New Zealand**: Although the adoption of AI is in the early stages in New Zealand, many food companies are already using it. Research on AI technology is taking place across all universities in New Zealand. New Zealand needs to engage with AI to shape a prosperous, inclusive, and thriving future for the country [27].

- **Africa**: In Africa, AI can help with some of the continent's most pervasive challenges: reducing poverty, improving education, delivering healthcare, eradicating diseases, addressing sustainability issues, and meeting the growing demand for food. Academic and research institutions across Africa are increasingly pursuing AI. For example, the University of Lagos recently launched the first AI Hub in Nigeria. Kenya was the first African country to launch an open-data portal to make information on education, energy, health, population, poverty, and water and sanitation. Microsoft created the doctoral scholarship program to support research collaborations between academics in the Europe, Middle East, and Africa (EMEA) region with researchers at the Microsoft Research Cambridge Laboratory [28].

11.8 FUTURE OF AI IN FOOD INDUSTRY

Artificial intelligence is poised to revolutionize the food and beverage industry, as AI-equipped tools can significantly increase efficiency without compromising on quality. Market trends depend on customer ideology towards the food or product and demand for health awareness and wellness. Quotes such as "Eat Global-Buy Local", "Organic = Healthy", "Gluten-free in case of major Asian markets" states the huge influence of health awareness and wellness on market trends. Here two major trends in food and beverage industry to watch out for in the coming years.

Automation: Automation has made its way into almost every industry. The food companies are already adopting AI to automate different parts of their operations. Within few decades, the food industry has attained peaks of productivity with the development of automation [24]. The automation systems use various technologies such as lasers, cameras, X-rays, and Near Infra-red spectroscopy to analyze each fruit or vegetable. Automation is being used than ever before. For example, automatic systems can collect hundreds of pieces of data on a single fruit in seconds and rapidly make an assessment about it. Robots can be used for seeding, spraying water, harvesting, processing, and packaging of food products. Robotics and automation in the food industry are growing with more and more applications. Robotics and automation in the food industry are growing with more and more applications.

Food Processing: The food processing industry is unique and complicated since there are so many products a single company can provide. Market trends shape the food processing technologies which will influence food industry eventually. In the coming five years, major food processors may be able to employ recommendation engines to suggest new products and flavor combinations to customers. AI is revolutionizing food processing from Farm to Fork. AI can be used more and more by food processing plants to improve feedstock sorting, find efficiencies, and increase safety. AI is reshaping the food processing business and will revolutionize the sector forever.

11.9 CONCLUSION

Artificial Intelligence (AI) is a way of making intelligent machines that think, work, and react like humans. It is the theory and development of computer systems that can perform tasks that normally require human intelligence. AI is pervading and transforming all aspects of the food and beverage industry. The implementation of artificial intelligence in the food industry is already moving the industry to new heights. While there is much promise in AI, the future depends on deeper engagement among food industry, policy makers, and technologists. AI-enabled technologies can be a possible tool in dealing with future demand for sustainable food supply.

Today, forward-looking businesses in the food industry are already adopting AI to automate different parts of their operations. Tech giants like Google, Microsoft, and IBM are highly interested in developing AI. The entire society can benefit from the use of AI in the food industry. It is time to address world hunger using AI and machine learning. More information on AI in the food industry can be found in the following related journals:

- Artificial Intelligence Review
- Artificial Intelligence in Agriculture
- AI Magazine
- British Food Journal
- Journal of Food Engineering
- Journal of Agriculture and Food Research
- International Journal of Food Properties
- Food Science and Technology

REFERENCES

[1] **M. N. O. Sadiku, T. J. Ashaolu,** and **S. M. Musa,**" Food industry: An introduction," International Journal of Trend in Scientific Research and Development, vol. 3, no. 4, May- Jun. 2019, pp. 128-130.

[2] **M. N. O. Sadiku, T. J. Ashaolu,** and **S. M. Musa,** "Emerging technologies in agriculture," International Journal of Scientific Advances, vol. 1, no. 1, July-August 2020, pp. 31-34.

[3] **H. Chen, L. Li,** and **Y. Chen,** "Explore success factors that impact artificial intelligence adoption on telecom industry in China," Journal of Management Analytics, 2020.

[4] **S. Greengard,** "What is artificial intelligence?" May 2019, https://www.datamation.com/artificial-intelligence/what-is-artificial-intelligence.html

[5] https://in.pinterest.com/pin/828662400161409072/

[6] **A. Kaplan** and M. **Haenlein**, "Rulers of the world, unite! The challenges and opportunities of artificial intelligence," Business Horizons, vol 63, no. 1, January–February 2020, pp. 37-50.

[7] **M. N. O. Sadiku, O. I. Fagbohungbe**, and S. **M. Musa**, "Artificial intelligence in food industry," International Journal of Engineering Research and Advanced Technology, vol. 6, no. 10, October 2020, pp. 12-19.

[8] **S. Kurilyak**, "Coronavirus update: Food and artificial intelligence," February 2019, https://blog.produvia.com/artificial-intelligence-ai-in-food-industry-ec8e925fa35e

[9] **K. Utermohlen**, "4 Applications of artificial intelligence in the food industry," January 2019,
https://heartbeat.fritz.ai/4-applications-of-artificial-intelligence-ai-in-the-food-industry-e742d7c02948#:~:text=1)%20Sorting%20Food,of%20canned%20and%20bagged%20goods.&text=One%20of%20the%20most%20advanced,solutions%20with%20machine%20learning%20functionalities.

[10] "Machine learning and artificial intelligence in the food industry," August 2019, https://medium.com/@spd.group/machine-learning-and-artificial-intelligence-in-the-food-industry-598f78471106

[11] "How artificial intelligence is revolutionizing the food and beverage industry," May 2020,
https://new.abb.com/news/detail/61772/how-artificial-intelligence-is-revolutionizing-the-food-and-beverage-industry#:~:text=AI%20can%20play%20a%20vital,complexity%20to%20the%20production%20line.

[12] "Applications of artificial intelligence (AI) in the food industry," https://morioh.com/p/6f346ea78169

[13]" Adoption of artificial intelligence in agriculture," *Bulletin UASVM Agriculture*, vol. 68, no. 1, 2011, pp. 284-293.

[14] **K. Jha et al.**, "A comprehensive review on automation in agriculture using artificial intelligence," Artificial Intelligence in Agriculture, vol 2, June 2019, pp. 1-12.

[15] **S. Francis**, "Opinion: How artificial intelligence can help the food industry," May 2018, https://roboticsandautomationnews.com/2018/05/23/opinion-how-artificial-intelligence-can-help-the-food-industry/17383/

[16] "What are the application of artificial intelligence use in the food processing industry?" https://shrutinearlearn.wixsite.com/mysite/post/what-are-the-application-of-artificial-intelligence-use-in-the-food-processing-industry

[17] **R. Patel**, "artificial intelligence ai in restaurant business – Benefits, possibilities & future," January 2020, https://aglowiditsolutions.com/blog/ai-in-restaurant-business/

[18] **J. Iqbal**, **Z. H. Khan**, and **A. Khalid**, "Prospects of robotics in food industry," *Food Science and Technology*, viol. 37, no. 2, April-June 2017, pp. 159-165.

[19] "Robotics and automation in the food industry and its future," December 2019, https://www.lacconveyors.co.uk/robotics-and-automation-in-the-food-industry/

[20] **J. Iqbal**, **Z. H. Khan**, and **A. Khalid**, "Prospects of robotics in food industry," *Food Science and Technology*, vol. 37, no. 2, April-June 2017, pp. 159-165.

[21] **A. Eliazàt**, "4 Ways AI is revolutionizing the food industry," June 2020, https://autome.me/4-ways-ai-is-revolutionizing-the-food-industry/

[22] **"5 Applications of ai in food processing industry,"** April 2020, https://usmsystems.com/artificial-intelligence-in-food-processing-industry/

[23] "Artificial intelligence applications in food industry," October 2019, https://www.linkedin.com/pulse/artificial-intelligence-applications-food-industry-mert-damlapinar

[24] **M. Damlapinar**, "Artificial intelligence applications in food industry," October 2019,

https://www.linkedin.com/pulse/artificial-intelligence-applications-food-industry-mert-damlapinar

[25] **V. Kakani et al.**, A critical review on computer vision and artificial intelligence in food industry," Journal of Agriculture and Food Research, vol. 2, December 2020.

[26] **A. Connolly**, "Artificial intelligence can save the food industry," August 2020, https://www.linkedin.com/pulse/can-artificial-intelligence-save-food-industry-aidan-connolly#:~:text=Reducing%20the%20presence%20of%20pathogens,limiting%20consumer%20illness%20or%20recalls.

[27] "Artificial intelligence: Shaping a future New Zealand," https://www.mbie.govt.nz/dmsdocument/5754-artificial-intelligence-shaping-a-future-new-zealand-pdf#:~:text=Our%20modelling%20analysis%20finds%20that,choices%2C%20but%20in%20unexpected%20ways.

[28] "Artificial intelligence for Africa: An opportunity for growth, development, and democratisation," https://www.up.ac.za/media/shared/7/ZP_Files/ai-for-africa.zp165664.pdf

CHAPTER 12
AI IN AUTONOMOUS VEHICLES

"Self-driving cars surely will make a huge contribution to society. We'll be able to redesign the urban environment so that parks will replace parking lots. Think of the money we'll save, the reduction in accidents and the incredible freedom this will provide people who can't drive today." - *Jen-Hsun Huang*.

12.1 INTRODUCTION

Mobility is a basic necessity of contemporary society, and it is ensured by a three-component system: vehicle, driver, and transport infrastructure. Mobility in emerging markets often faces acute challenges due to poor infrastructure, growing populations, urbanization, and pollution. Globally, approximately 1.25 billion road traffic deaths occur every year. Engineers and scientists all over the globe are offering solutions and struggling to make human life more comfortable. One of their most spectacular recent developments is autonomous vehicle (AV). Autonomous vehicles have been hitting the headlines lately. They usually refer to self-driving vehicles that can fulfill transportation capabilities of a traditional vehicle. They are not designed to obey the law but for the purposes of promoting safe and efficient traffic and driving from point A to point B [1]. They are regarded as a post-Uber disruption to public commute and transportation of goods. They are starting to become a real possibility in main economic sectors such as transportation, agriculture, and military. AV technology is regarded as a significant market disruptor for multiple industries [2].

Autonomous (also called self-driving, driverless or robotic) vehicles are the future of automobiles. The autonomous vehicle segment is the fastest growing segment in the automotive industry. For this reason, a host of auto heavyweights are investing in autonomous R&D and developing self-driving vehicles. Such companies include Amazon, Audi, Apple, Waymo, Tesla, BMW, Google, General Motors, Volkswagen, Volvo, Ford Motor, IBM, Microsoft, Mercedes-Benz, Bosch, Nissan, Alphabet, Honda Motor, and Uber Technologies.

Autonomous vehicles have been in development for almost thirty years. For a vehicle to be autonomous, it must be continuously aware of its surroundings, by perceiving and then acting on the information. Autonomous vehicle (AV) is one of the key application areas of artificial intelligence (AI).
In other words, AI is the most important component of autonomous and connected vehicles. This is due to the humongous amount of environmental data generated by the

vehicle. AVs depend on AI to interpret the environment, understand its conditions, and make driving-related decisions. Autonomous vehicles employ numerous cameras, sensors, radars, communication systems, and AI technology to enable the vehicle to generate massive amounts of data which can be processed in fractions of a second to help the vehicle to see, hear, think, and make decisions just like human drivers do [3].

This chapter presents uses of AI technology in autonomous vehicles. It begins by briefly review AI. It explains the concept of an autonomous vehicle. It presents the roles of AI and machine language in autonomous vehicles. It covers some specific applications of AI in autonomous vehicles. It highlights the benefits and challenges of AV. It describes how AVs are implemented worldwide. It addresses the future of AI in AVs. The last section concludes with comments.

12.2 REVIEW OF ARTIFICIAL INTELLIGENCE

Artificial intelligence (AI) is the foundation of autonomous vehicles (AVs). AI refers to computer systems that mimic human cognitive functions. It is a field of computer science that deals with intelligent machines. The term "artificial intelligence" was first used at a Dartmouth College conference in 1956. The main goal of AI is to enable machines to perform complex tasks that typically require human intelligence [4]. In simple terms, AI attempts to clone human behavior. An important feature of AI technology is that is can be added to existing technologies. AI is now one of the most important global issues of the 21st century. It is poised to disrupt our world and change processes and developments in the field of science, engineering, education, business, and agriculture [5]. Figure 12.1 shows some artificial intelligence applications.

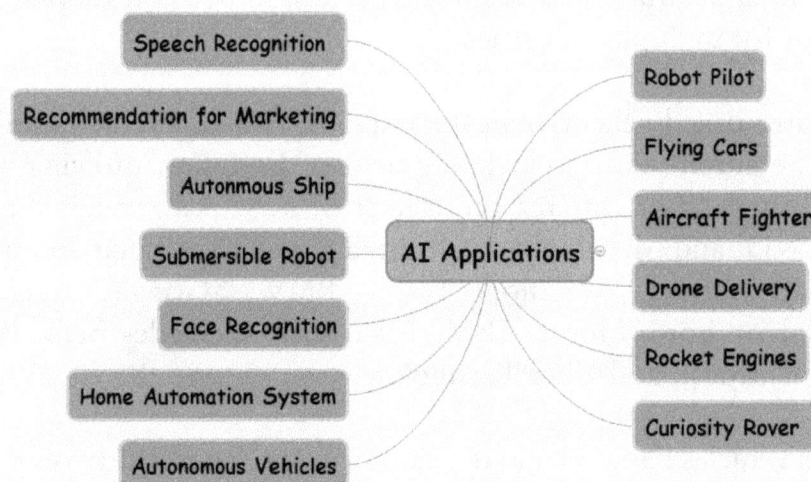

Figure 12.1 Artificial intelligence applications.

The concept of AI is an umbrella term that encompasses many different technologies. AI is not a single technology but a collection of techniques that enables computer systems to perform tasks that would otherwise require human intelligence [6]. The major disciplines in AI include [7]:

- Expert systems
- Fuzzy logic
- Neural networks
- Machine learning
- Deep learning
- Natural Language Processors
- Robots

These AI tools are illustrated in Figure 12.2.

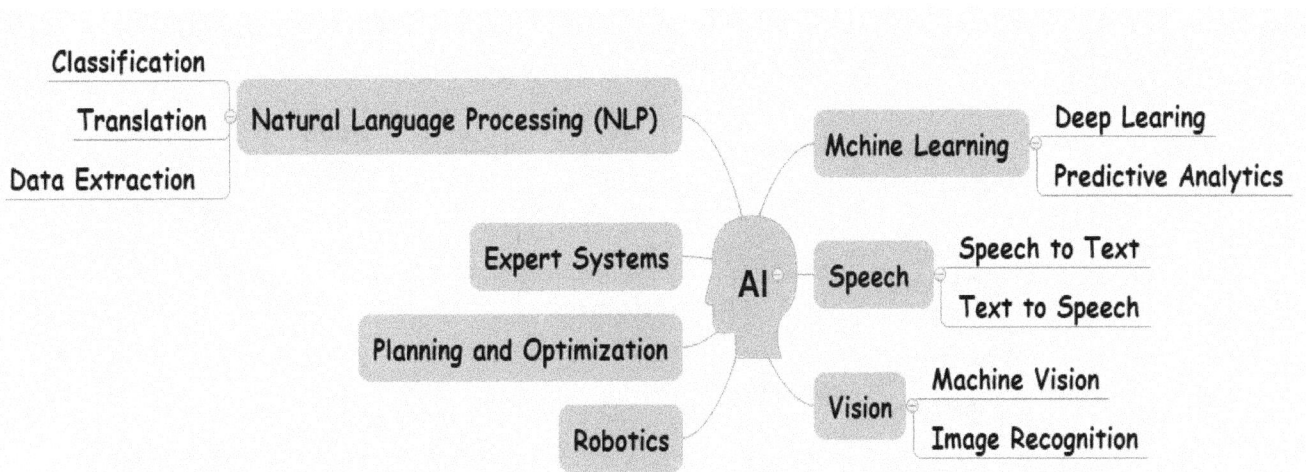

Figure 12.2 Branches of Artificial intelligence.

Each AI tool has its own advantages. Using a combination of these models, rather than a single model, is recommended. AI systems are designed to make decisions using real-time data. AI algorithms, such as machine learning and deep learning, are most commonly used to make intelligent predictions. They have the ability to learn and adapt as they make decisions. AI is no longer just an academic field; machine learning and deep learning are becoming mainstream technologies that manufacturing companies can harness. AI tools have become an absolute necessity to make autonomous vehicles function properly and safely.

12.3 CONCEPT OF AUTONOMOUS VEHICLES

Autonomous vehicles constitute one of the most spectacular recent developments of AI. As opposed to human-driven vehicles, autonomous vehicles essentially refer to self-driving vehicles. They are smart vehicles that are able to perceive their environment and to move on accordingly without human intervention. They operate with the capability to have automatic motions and navigate themselves depending on the environments and scheduled tasks [8]. Figure 12.3 shows the architecture of autonomous car.

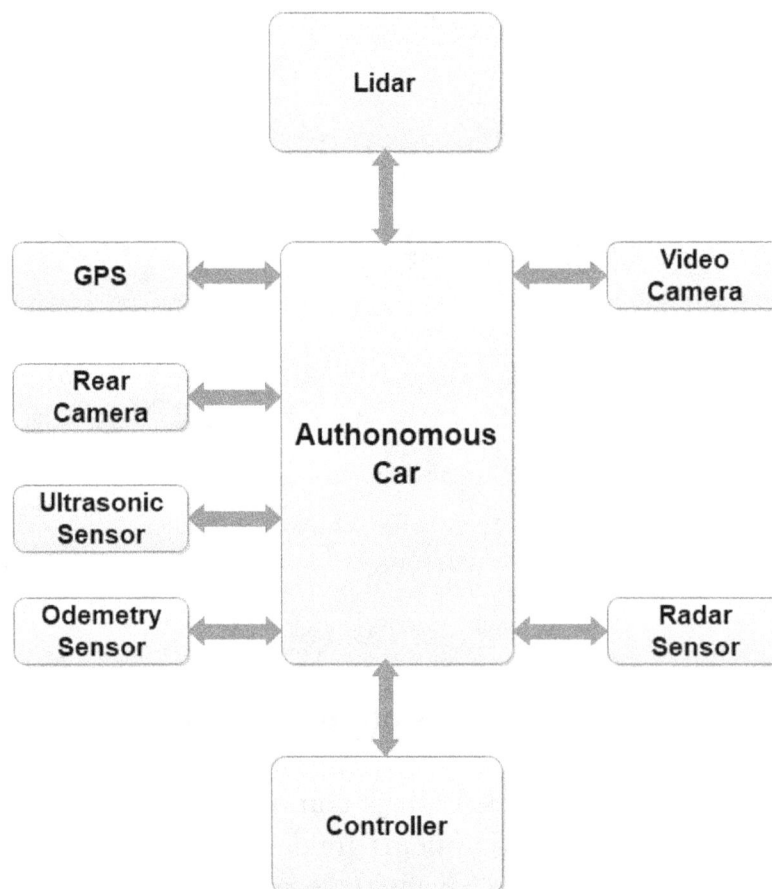

Figure 12.3   Architecture of autonomous car.

An AV or driverless car is an ambitious project which requires the fusion of many technologies like electronics, communications, mechatronics, software engineering, artificial intelligence, GPS, and industrial IoT. It is a vehicle that uses a combination of sensors, cameras, radar, and AI to travel between destinations without a human operator.

It is designed to be able to detect objects on the road, maneuver through the traffic without human intervention, and get to the destination safely.

It is fitted with AI-based functional systems such as voice and speech recognition, gesture controls, eye tracking, and other driving monitoring systems. Several companies have announced their plan to get involved in autonomous or driverless and electric vehicle technology.

Connected and AVs are now becoming a cornerstone of the increasingly connected world. They are receiving a lot of attention from manufacturers, service providers, governments, universities, consumers, and other stakeholders. The main goal of autonomous vehicles is to build a self-driving system that can perceive the road better than the best human driver. They are an incredible innovation that will likely transform transportation, especially in urban environments, in the near future. Although autonomous vehicles can improve performance and safety, there are a myriad of serious technology, regulatory, and security challenges to consider in preparation for full vehicle autonomy [9]. A typical example of autonomous vehicle is shown in Figure 12.4.

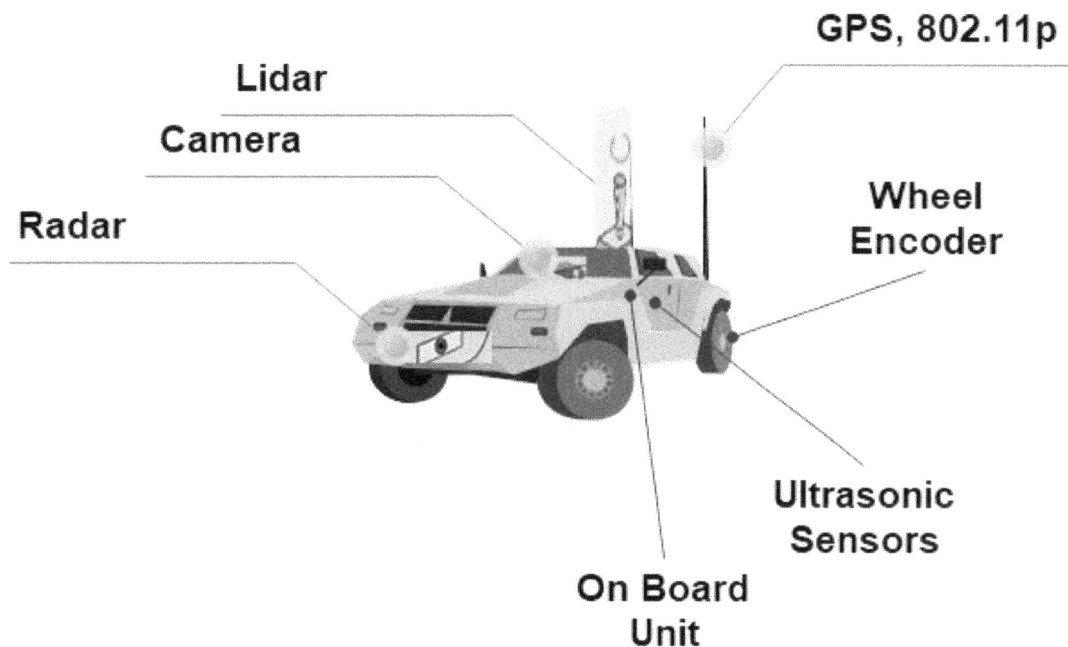

Figure 12.4 Autonomous/connected vehicle.

Autonomous vehicles combine AI and robotics. They are regarded as a promising answer to traffic jams, accidents, and environmental pollution. They will constitute the backbone of future next-generation intelligent transportation systems (ITS) providing travel comfort and road safety along with a number of value-added services. They are used in search and rescue, urban reconnaissance, mine detonation, supply convoys, etc. [10]. They can help save lives on the battlefield.

Autonomous Vehicle (AV) is also described as "driverless vehicle," "robotic vehicle," or "self-driving vehicle." AV is regarded as a multidisciplinary technology. The enabling technologies in support of connected autonomous vehicles include camera, GPS and GNSS, and sensors, radar, LiDAR (Light Detection and Ranging), and Internet of things. The race to develop autonomous vehicles has heated up with many major automotive manufacturers such as Tesla, Audi, General Motors, Mercedes Benz, Uber, Google, and Amazon.

SAE International (formerly the Society of Automotive Engineers) classifies AVs on a scale of 0 to 5. The six levels are presented as follows [11,12]:

Level 0: No automation: All driving tasks and major systems are controlled by a human driver. The automated system has no vehicle control but can issue warnings.

Level 1: Function-specific automation: Provides limited driver assistance. The driver must be ready to take control at any time.

Level 2: Partial driving automation: At least two primary functions are combined to perform an action. The driver is obliged to detect objects and events and react if the automated system does not respond correctly.

Level 3: Conditional driving automation: Enables limited self-driving automation. Vehicles at this level can make informed decisions for themselves. In known environments (such as highways), the driver can safely divert his attention from driving tasks.

Level 4: High driving automation: An automated driving system performs all dynamic tasks of driving. The automated system can control the vehicle in almost any environment, such as extreme weather conditions, and fewer parking spaces.

Level 5: Self-driving automation: An automated driving system performs all dynamic functions of driving. No human intervention is required. A vehicle at this level requires no driver. It is on its own and must be able to react to all situations that might arise.

The six levels are shown in Figure 12.5 [13] and are summarized as follows: No Automation, Driver Assistance, Partial Automation, Conditional Automation, High Automation, and Complete Automation.

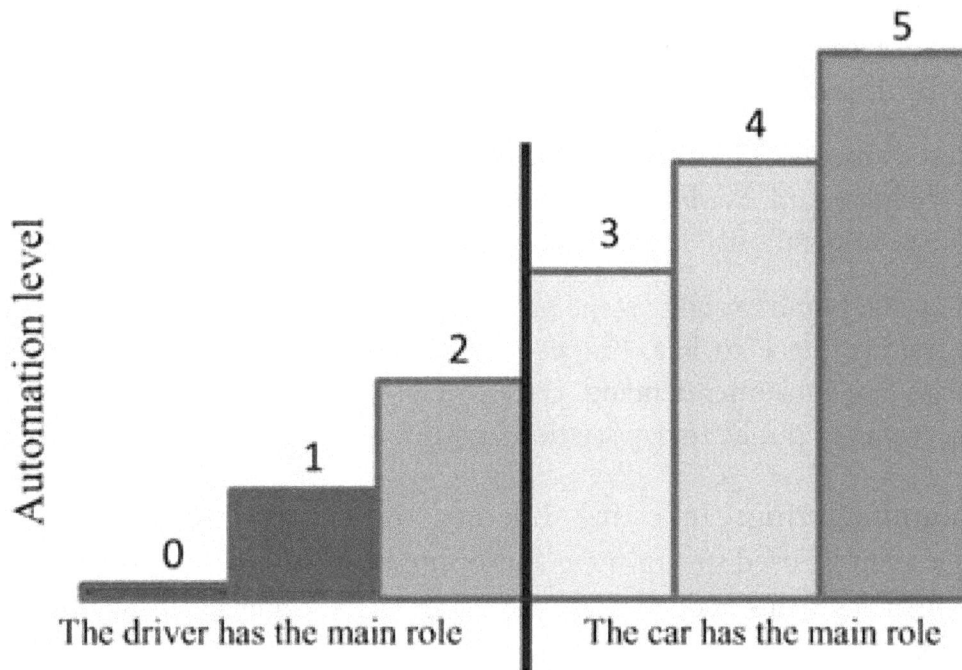

Figure 12.5 The six levels of autonomy [13]

The classification has been adopted by DOT. Vehicles sold today are in levels 1 and 2. Levels 3, 4, and 5 will probably increase vehicle prices significantly. But how do we get to Level 5?

12.4 AI IN AUTONOMOUS VEHICLES

AI is a branch of computer science that deals with anything related to making machines smart. AI uses data and algorithms to replicate human decision or thinking ability. For AI to work in AV, it needs IoT devices such as radars, ultrasound, cameras, LiDAR, accelerometers, and gyroscopes. AI performs several tasks in a self-driving automobile. One of the main tasks is path planning, which involves the navigation system of the vehicle. Path planning is needed to optimize the trajectory of the vehicle and to lead to better traffic patterns. AI helps vehicles to navigate through the traffic and handle complex situations. AI can respond quickly to data points generated from hundreds of different sensors. Sensors provide information for the road, other vehicles about the road, and other impediments.

AI in the auto industry can be classified into four segments [14]:

Autonomous driving: AVs or self-driving vehicles are becoming more and more desirable. AVs can bring safety because they are much more alert and will not be overcome by distractions.

Connected vehicles: Vehicles are quickly transforming into connected devices. Cars now use cellular and Wi-Fi connections to upload and download entertainment, navigation, and operational data. AI is an essential technology for connected vehicles.

Mobility as a service: Mobility is the lifeblood of any city. It is what makes a city livable and an attractive place to live. Car companies are becoming mobility companies to address changing consumer demand. Car ownership in urban areas may decline in favor of various forms of public transportation and ride sharing.

Smart manufacturing: Industrial Internet of Things (IIoT) and Industry 4.0 technologies can be used to automate and optimize manufacturing processes. The use of AI in vehicles helps reduce manufacturing costs while ensuring safer and more innovative vehicle production. Car manufacturers are using AI tools in every aspect of the car-making process.

12.5 MACHINE LEARNING IN AUTONOMOUS VEHICLES

Machine learning (ML) is a major approach to realize AI. It is one of the core technologies used in autonomous vehicles. It refers to the process by which computers learn from data so they can perform some specific tasks. ML algorithms can be classified as a supervised algorithm and an unsupervised algorithm. A major task of a machine learning algorithm in AVs is continuous rendering of the surrounding environment and the prediction of possible changes to those surroundings. For safe navigation, AVs require the ability to make decisions. Machine learning algorithms enable AVs to make decisions in real time. They can be successfully and reliably used to perform virtually all mobility functions such as navigation, safety maneuvers, entertainment, and parking. Machine learning is being deployed for the perception and understanding of the world around the vehicle. This involves the use of camera-based systems to detect and classify objects. This machine learning-based application can also incorporate the driver's gesture and speech recognition, and language translation.

Deep learning technology, which is a technique for implementing machine learning, is expected to be the fastest-growing technology in the automotive AI market.

12.6 APPLICATIONS OF AI IN AUTONOMOUS VEHICLES

AI is revolutionizing the transportation sector. Successful applications of AV technologies could fundamentally transform various industries such as automotive, transportation, energy, agriculture, battlefield, space, the deep sea, and other dangerous environments. Autonomous vehicles include deep space probes, spacecraft, unmanned aerial vehicles (UAV), unmanned ground vehicles (UGV), unmanned sea/surface vehicles (USV), and unmanned underwater vehicles (UUV) [15].

- **Autonomous cars and other highway vehicles:** The automotive industry is focusing on the integration of AI in autonomous cars, while other application areas include R&D, procurement, supply chain management, manufacturing, mobility services, and customer experience. Cars are at the top of the list of autonomous vehicles. Figure 12.6 illustrates a man in an autonomous vehicle [16].

Figure 12.6 A man in an autonomous vehicle [16].

In 2015, Amazon filed for a patent for a system that helps autonomous cars navigate roadways. Traditionally, automotive manufacturing was dominated by internal combustion engine. Today, car manufacturers are implementing a range of AI technologies to mimic, augment, and support actions of humans, including voice controls, telematics, interior-facing cameras, touch-sensitive surfaces, and personalized platforms. Autonomous trucks have been tested in the U.S. and Europe to enable drivers to use autopilot over long distances.

- **Autonomous flight vehicle:** The integration of AI with the aircraft that could automate the task of manual flight and navigation would be considered a milestone in the field of aeronautics. The idea of autonomous flight vehicle incorporating AI on a large scale involves the commercial and military aircrafts and drones autonomously operating with little or no harm interaction. Such autonomous vehicles will execute shortest and fastest flight routes with more efficient path planning, which in turn will reduce the fuel consumption [17]. Large aircraft have been using automatic landing systems for several years.

- **Autonomous undersea vehicle:** They have also been called autonomous surface craft. As global positioning systems (GPS) have become more compact, effective, and affordable, AUVs have become more capable.

- **Autonomous agricultural vehicles:** These are characterized as the autonomous guidance system for agricultural vehicles including navigation sensors (GPS, machine vision, sensors, etc.), computational methods for features and fuse extraction, navigation planners to supply control algorithms, and steering controllers. AI and robotic technology will be key tools for such vehicles [18].

Other applications of AI in the automotive industry include predictive maintenance and driver assistance.

12.7 BENEFITS

Autonomous vehicles have been receiving a lot of attention recently largely due to the technological boom of AI. They promise to reduce traffic congestion and collisions, reduce accident and pollution problems, improve traffic flow, mobility, relieve individuals from driving, decrease fuel consumption, and facilitate transportation and businesses operations.

Based on the data collected over time, AI can speculate and provide preferences such as seat position adjustment, screen adjustment, controls air input, and songs to be played.

Other benefits include [19]:

- **Safety and privacy:** Security and privacy are always being the major issue associated with the electronic systems. The main benefit of autonomous vehicles is safety. AVs are fully capable of sensing the surrounding environment to enhance roadway safety. AVs are designed to choose the course of action that causes the less damage to everybody. Autonomous cars can remove risk factors from the equation and significantly reduce the number of accidents or crashes. AI techniques are also being implemented in vehicles to provide more safety by simply removing the need of a human driver, the most common cause of traffic accidents. To ensure safety, automakers are introducing many exciting features such as automatic emergency braking, lane crossing alerts, driver and passenger monitoring, blind spot elimination, side collision warning, and optional self-driving capabilities. Risk of cyberattack will increase since the AVs are a moving computer.

- **Consumer behavior:** To be successful, AVs have to cope with individual idiosyncrasies and variation in needs. Deployment of autonomous vehicles on public roads requires understanding the intent of human drivers and adapting to their driving styles. Interacting with human drivers is a major challenge of autonomous driving. Changes in consumer behavior and technology are disrupting traditional modes of operation. AI and perception technologies promise to provide a safer and more deterministic behavior which will lead to benefits such as fuel efficiency, comfort, and convenience.

- **Smart mobility:** AVs are an essential part of smart mobility in smart cities. They will increase the mobility of individuals who do not drive as a result of age or disability. AI promotes advanced driving so that people can experience easy navigation. AI can help provide customized entertainment while the vehicle is traveling.

- **Ease of claim:** Data from the vehicle can be used for faster processing of claims in case of accidents. This can contribute to decrease in prices for insurance, since the safety is more deterministic and guaranteed.

- **Less equipment failure:** If a machine fails unexpectedly on an automotive assembly line, the costs can be catastrophic. AI-based algorithms can digest masses of data from sources, detect anomalies, separate errors from background noise, diagnose the problem, and predict if a breakdown is likely or imminent.

- **Quality control:** Once the parts are put into production, it is necessary to carefully control their quality. Sensor-based AI can assess the quality of every part on the production line and detect defects more accurately than humans.

- **Leaner supply chain:** Automotive supply chains are complex networks. AI-powered supply chains are being used to analyze a massive amount of data to be able to forecast accurately. AI-powered supply chains will allow fully automated self-adjusting systems to make supply-chain management decisions autonomously, adjusting routes and volumes to meet predicted demand spikes.

- **Predictive maintenance:** Predictive analytics is one of the strongest capabilities of AI and machine learning. The essence of predictive maintenance is that the system analyzes the equipment, compares its specs with industry and safety standards. The essence of reactive maintenance is to replace a critical part before it crashes the production system.

12.8 CHALLENGES

There are still several aspects of autonomous vehicles that are presenting significant challenges and some of the challenges are due to the intrinsic nature of AI. This section presents specific challenges for developing, testing, and deploying AI technologies in the AV. Some serious obstacles such as adequate regulations and international standards for digital infrastructure remain.

Other challenges include [20]:

- **Complexity:** Autonomous driving systems are becoming increasingly complex and must be tested effectively before deployment. This necessitates long and numerous rounds of testing. Intensive AI algorithms consume more power, especially for electric vehicles. Accidents with testing vehicles often happen because the simulation environment is different from the real-world conditions. The autonomous vehicle will contain more lines of code than any other software platform that has ever been created.

- **Cost:** Manufacturers of AVs have spent heavily in building those vehicles. For example, Google paid about $80,000 for the AV module which is way out of a normal man's reach. It is hoped that in the future, the prices of AVs will come close to the cost of traditional vehicles.

- **No regulation and standard:** There are no clear regulations on data collection and governing the new methods of autonomous transportation. Lawmakers and regulators are yet to determine who is liable when an autonomous vehicle is involved in an accident. AV technology presents a huge challenge to standardization and legal bodies. Governments have a major role to play in legislating the autonomy of vehicles and the absence of a driver.

- **Reliability:** Both software reliability and hardware reliability are possible, but quite expensive. The software must be right the first day the car is in service. The cost and time needed to provide completely bug-free software will be the end of the driverless car dream.

- **No user's trust:** Automakers are facing the need of gaining the user's trust. The willingness of people to trust AI is increasing at a very slow rate. Modern users want a car to be functional, comfortable, and safe.

- **Amount of data:** AI techniques need to interact with numerous sensors and use data in real time. This is related to achieving safety while driving the vehicle. Great amount of data and expertise is required to build and leverage simulators. To mimic human behaviors is difficult. The autonomous driving task as a whole becomes hugely complex due the massive amount of data involved.

- **Social dilemma:** Autonomous vehicles are meant to reduce traffic accidents, but they will sometimes have to choose between two evils, such as running over pedestrians or sacrificing themselves and their passenger to save the pedestrians. Developing the AI algorithms that will help AVs to handle this social dilemma is formidable. Policy makers need Solomon's wisdom to solve such paradoxes. AI technology is behind this superior spiritual intelligence.

- **Cybersecurity:** Security and privacy are always being the biggest issue associated with the electronic system. The cybersecurity of AVs is usually approached through the angle of the security of digital systems. The AVs are vulnerable to malicious attacks. Autonomous cars are based on the AI system, where it also requires using the Internet, which is a medium that can easily be abused by the hackers.

All of these challenges explain the slower adoption of AI in the automotive industry. They must be addressed by data engineers, computer scientists, and AI scientists, and other researchers to achieve full autonomy, a state in which the vehicle can navigate itself through any terrain in any conditions at any time.

12.9 GLOBAL USE OF AI IN AUTONOMOUS VEHICLES

Autonomous vehicles are becoming a real possibility in some sectors such as agriculture, transportation, and military. Here we consider the applications of AI in autonomous vehicles in different nations.

- **United States:** The United States is presently the largest automobile market in the world. Vehicles with crash-warning systems, lane-keeping framework, adaptive cruise control system, anti-lock braking systems, automatically activated safety mechanisms, and self-parking technology are already in use in the United States. Americans spend over 290 hours per year driving, making it one of the most common (and dangerous) points of human-machine interaction. The National Highway Traffic Safety Administration (NHTSA)[2] contends that automated vehicles can reduce injuries. In early 2015, Ford announced its "Smart Mobility Plan" to push the company forward in innovative areas including vehicle connectivity and autonomous cars. As part of its 10-year AV plan, Ford is steadily increasing its fleet and currently has around 100 autonomous test vehicles. It has pioneered the testing of self-driving cars in environments including snowy weather and complete darkness.

The Covid-19 pandemics has forced Ford to delay the launch of its self-driving service to 2022. GM planned to deploy thousands of self-driving electric vehicles beginning in 2018. A full commercial self-driving operation is taking longer to materialize than GM initially hoped, partly due to Covid-19 crisis. Microsoft is focusing its autonomous vehicle efforts [21].

- **The Netherlands:** Dutch traffic law poses some challenges for fully autonomous self-driving cars. One challenge is making the behavior of AVs conform to the traffic laws. The safety and efficiency purposes of traffic law require it to be precise. Dutch traffic laws that are easy for humans seem very hard for the current autonomous vehicles. For example, while the Google car may avoid colliding with a police officer, it may fail to obey the officer's directions [22].

- **China:** The expansion into China is a major strategic move for Aptiv, a self-driving software company. China is expected to account for more than two-thirds of autonomous miles driven worldwide by 2040. FAW Group, a China-based manufacturer of trucks, buses, and cars, is moving forward with several self-driving initiatives.

- **Israel:** Israel is evolving into a major hub for self-driving technology, with Intel, Continental, Samsung, Daimler, and General Motors also making investments or setting up shop in the country.

- **European Union:** In 2019, the European Union's Commissioner for Transport has said she expected fully autonomous driving capabilities to arrive by 2030. The European Telecommunications Standards Institute (ETSI) has developed a set of technical specifications to define an Intelligent Transport System (ITS) security architecture. French president responded sharply to rising pressure from the industrial sector and plans to grant driving licenses for level 3 cars , country-wide, from 2020. Fully autonomous vehicles are to be tested in France very soon [23].

12.10 THE FUTURE OF AI IN AUTONOMOUS VEHICLES

The AV is but one application of AI technologies which has a significant bearing on contemporary and future society [24]. Future autonomous (electric) vehicles are essentially software-driven compared to conventional vehicles.

Cars are being manufactured all over the world, with each manufacturer in intense competition with one another to produce the best vehicle. Autonomous cars are the future smart cars which are expected to be driverless, efficient, crash avoiding, and the ideal urban car of the future. Some are working tirelessly to create their very own self-driving vehicle from scratch. AI is a critical technology required for realizing autonomous driving. Car manufacturers around the globe are using AI in just about every facet of the car manufacturing process. AI is changing the way cars are manufactured globally. Due to the various challenges of AI in autonomous vehicles, barriers to widespread adoption remain. In the near future, AI will enable autonomous vehicles to become mainstream. Technology companies are at the forefront, leveraging their AI experience to capture the autonomous vehicle market.

Connected and automated vehicle has become the focal point of current transportation studies (covering topics like automation, car visions systems, and AI) and has a crucial role to play in the future of transportation. The demand and the need for autonomous vehicle technology is almost there. As the AV technology matures, personal and public transportation will be greatly transformed. A day is fast approaching when you can commute to work with driverless car, without needing to watch the road [25]. Figure 12.7 shows seven trends shaping the automotive industry by 2030.

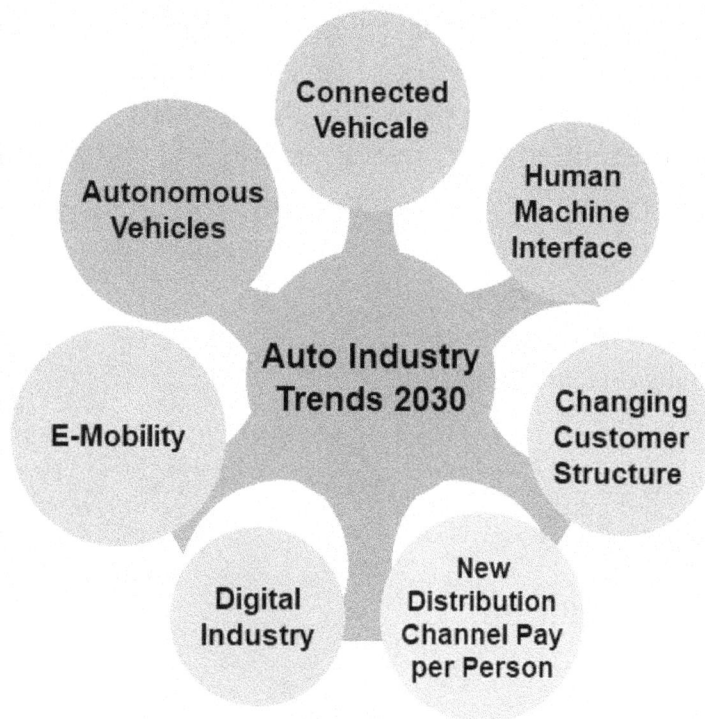

Figure 12.7 Seven trends shaping the automotive industry by 2030.

12.11 CONCLUSION

Artificial intelligence is making inroads in the automotive industry. It is the backbone of self-driving, autonomous or connected vehicles. It is being harnessed to bring autonomous driving from fantasy to reality. Autonomous cars are essentially the future smart cars anticipated to be driverless, efficient, crash avoiding, and ideal car of the future.

AVs must be installed on a mobile platform. Leading automakers use the AI technologies in their operations from design development to the sale of a car. Driverless vehicles (cars and trucks) are certainly on the horizon. To stay on top of the developments in AI in autonomous vehicles, one should consult the books in [26-36] and related journals:

- Artificial Intelligence Review
- Applied Artificial Intelligence
- Artificial Intelligence and Law
- AI Magazine

REFERENCES

[1] **H. Prakken**, "On the problem of making autonomous vehicles conform to traffic law," Artificial Intelligence and Law, vol 25, no. 3, pp. 341–363.

[2] **C. Medrano-Berumen** and **M. I. Akbaş**, "Validation of decision-making in artificial intelligence-based autonomous vehicles," Journal of Information and Telecommunication, 2020.

[3] M. **N. O. Sadiku**, **S. M. Musa**, and **A. Ajayi-Majebi**, "Artificial Intelligence in Autonomous Vehicles," International Journal of Trend in Scientific Research and Development, vol. 5, no. 2, Jan.-Feb. 2021, pp. 715-720.

[4] **S. Greengard**, "What is artificial intelligence?" May 2019, https://www.datamation.com/artificial-intelligence/what-is-artificial-intelligence.html

[5] https://medium.datadriveninvestor.com/artificial-intelligence-applications-space-to-underwater-6b0391fe8b2c

[6] https://in.pinterest.com/pin/828662400161409072/

[7] **S. Greengard**, "What is artificial intelligence?" May 2019, https://www.datamation.com/artificial-intelligence/what-is-artificial-intelligence.html

[8] **T. Raviteja** and **R. Vedaraj**, "An introduction of autonomous vehicles and a brief survey," Journal of Critical Reviews, vol 7, no. 13, 2020, pp. 196-202.

[9] **R. Elezaj**, "How AI is paving the way for autonomous cars," October 2019, https://www.machinedesign.com/mechanical-motion-systems/article/21838234/how-ai-is-paving-the-way-for-autonomous-cars

[10] **J. Connelly** et al., "Current challenges in autonomous vehicle development," Proceedings of. SPIE, May 2006.

[11] **S. Meryem** and T. **Mazri**, "Security study and challenges of connected autonomous vehicles," Proceedings of the 4th International Conference on Smart City Applications, October 2019, pp. 1–4.

[12] **A. Qayyum et al.**, "Securing connected & autonomous vehicles: Challenges posed by adversarial machine learning and the way forward," https://arxiv.org/pdf/1905.12762.pdf

[13] **I. Barabás**, "Current challenges in autonomous driving," Materials Science and Engineering, vol. 252, 2017, pp.

[14] **S. Rao**, "Artificial intelligence in the automotive industry," June 2010, https://blog.netapp.com/artificial-intelligence-in-the-automotive-industry/

[15] E. **Yagdereli**, **C. Gemci**, and **A. Z. Aktas**, "A study on cyber-security of autonomous and unmanned vehicles," Journal of Defense Modeling and Simulation: Applications, Methodology, Technology, 2015, vol. 12, no. 4, 2015, pp. 369–381.

[16] "A man in a autonomous driving test vehicle," https://www.123rf.com/stock-photo/autonomous_car.html?sti=njm1zmuighvkuev2le|&mediapopup=36488489

[17] **J. Makadia et al**, "Autonomous flight vehicle incorporating artificial intelligence," Proceedings of the International Conference on Computational Performance Evaluation, Meghalaya, India. July 2-4, 2020.

[18] **A. Roshanianfard et al.**, "A review of autonomous agricultural vehicles," Journal of Terramechanics, vol. 91, 2020, pp. 155–183.

[19] **M. Anirudh**, "AI & automotive — 8 Disruptive use-cases," March 2020, https://unfoldlabs.medium.com/ai-automotive-8-disruptive-use-cases-fd079926aea9

[20] **O. Odukha**, "Adoption of AI in the automotive industry: is it worth the effort?" February 2020,

https://www.intellias.com/adoption-of-ai-in-the-automotive-industry-is-it-worth-the-effort/#:~:text=Also%2C%20artificial%20intelligence%20in%20car,algorithms%20is%20worth%20the%20money

[21] "40+ Corporations working on autonomous vehicles," December 2020, http://groupementadas.canalblog.com/archives/2020/04/26/38226172.html

[22] **H. Prakken**, "On the problem of making autonomous vehicles conform to traffic law," Artificial Intelligence and Law, vol. 25, August 2017, pp. 341–363,

[23] **K. R. Ahmadi**, "Artificial intelligence and autonomous vehicles." https://drivetribe.com/p/artificial-intelligence-and-autonomous-SfevTlYUR0OFm97L41Bc6A?iid=H1EmRf-oRtau4o9tAiYlsA

[24] **M. Cunneen**, **M. Mullins**, and **F. Murphy**, "Autonomous vehicles and embedded artificial intelligence: The challenges of framing machine driving decisions," Applied Artificial Intelligence, vol. 33, no. 8, 2019, pp. 706-731.

[25] **O. Wyman**, "Future automotive industry structure — FAST 2030," Unknown source

[26] **L. B. Eliot**, New Advances in AI Autonomous Driverless Self-Driving Cars: Artificial Intelligence and Machine Learning. LBE Press Publishing, 2017.

[27] **S. Liu et al.**, Creating Autonomous Vehicle Systems. Morgan & Claypool; 2nd edition, 2020.

[28] **A. Raymond**, How Autonomous Vehicles Will Change the World: Why Self-Driving Car Technology Will Usher in a New Age of Prosperity and Disruption. Clever Books, 2020.

[29] **M. E. McGrath**, Autonomous Vehicles: Opportunities, Strategies and Disruptions.
Independently published, 2019.

[30] **P. Lin, K. Abney,** and **R. Jenkins (eds.**), Robot Ethics 2.0: From Autonomous Cars To Artificial Intelligence. New York: Oxford University Press, 2017.

[31] **M. Maurer et al. (eds.)**, Autonomous Driving: Technical, Legal and Social Aspects. Springer, 2016.

[32] **L. Eliot, AI Self-Driving Cars Divulgement**: Practical Advances in Artificial Intelligence and Machine Learning. LBE Press Publishing, 2020.

[33] **L. Eliot**, The Next Wave AI Self-Driving Cars: Practical Innovations in AI and Machine Learning. LBE Press Publishing, 2018.

[34] **S. Ranjan** and **S. Senthamilarasu**, Applied Deep Learning and Computer Vision for Self-Driving Cars: Build Autonomous Vehicles Using Deep Neural Networks And Behavior-Cloning Techniques. Packt Publishing, 2020.

[35] **I. J. Cox** and **G. T. Wilfong (eds.)**, Autonomous Robot Vehicles. Springer-Verlag, 2012.

[36] **P. Lin, K. Abney**, and **R. Jenkins**, Robot Ethics 2.0: From Autonomous Cars to Artificial Intelligence. New York: Oxford University Press, 2017.

CHAPTER 13
AI IN CHATBOTS

"Chatbots are not just software in the modern era. Chatbots are like our personal assistants who understand us and can be micro configured. They remember our likes and dislikes and never tend to disappoint us by forgetting what we taught them already, and this is the reason why everyone loves chatbot." - *Sumit Raj*

13.1 INTRODUCTION

Technology plays a major role in the modern living. As technology seeks to provide solutions for digital transformation, customer intelligence and machine intelligence are becoming critical enabling factors.

We are in the digital age, where the intelligent machines are all around us. Tasks that were once performed only by humans are now optimized, automated, and completed significantly faster by machines. There are number of new kinds of computer machines and software suites available which are collectively known as chatbots or bots. A chatbot is a software system that allows humans to interact with technology using a variety of input methods. It may also be regarded as a service that allows people interact with via a chat interface. It is a virtual assistant capable of answering a certain number of questions from human users, providing the correct answers [1]. Figure 13.1 shows a typical chatbot icon.

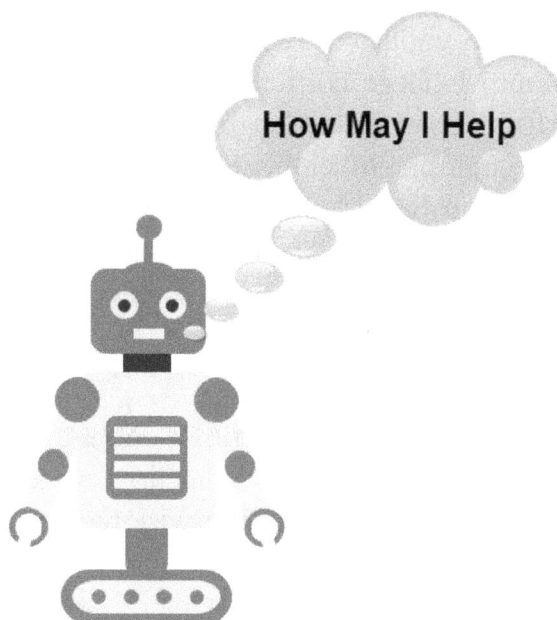

Figure 13.1 A typical chatbot icon.

Chatbots are currently used in various online applications [2]. They are now taking pizza orders, reserving hotel rooms, booking airflight, and scheduling doctor appointments. In short, these robots are becoming pervasive. It is quite possible that you have spoken to a chatbot without realizing it.

Chatbots (also known as conversational agents or virtual assistants) are the latest industry's newest tools designed to simplify the interaction between humans and computer machines. Since chatbots have entered the digital world, marketers are curious to use them as a major tool to interact with their customers on daily basis. Chatbots are already part of virtual assistants, such as Siri, Alexa, Cortana, and Watson Assistant.

This chapter provides an introduction on artificial intelligence-based chatbots. It begins by briefly reviewing AI. It then describes what chatbot is all about and how it works. It presents some notable chatbots. It covers some applications of AI-based chatbots. It highlights some benefits and challenges of AI chatbots. It addresses the future of AI chatbots. The last section concludes with comments.

13.2 REVIEW ON ARTIFICIAL INTELLIGENCE

Artificial intelligence (AI) is the foundation of the fourth industrial revolution. AI refers to computer systems that mimic human cognitive functions. It is a field of computer science that deals with intelligent machines. The term "artificial intelligence" was first used at a Dartmouth College conference in 1956. The main goal of AI is to enable machines to perform complex tasks that typically require human intelligence [3]. In simple terms, AI attempts to clone human behavior. An important feature of AI technology is that is can be added to existing technologies. AI is now one of the most important global issues of the 21st century. It is poised to disrupt our world and change processes and developments in the field of science, engineering, education, business, and agriculture.

The concept of AI is an umbrella term that encompasses many different technologies. AI is not a single technology but a collection of techniques that enables computer systems to perform tasks that would otherwise require human intelligence [4].

The major disciplines in AI include [5]:

- Expert systems
- Fuzzy logic
- Neural networks
- Machine learning (ML)
- Deep learning
- Natural Language Processors (NLP)
- Robots

These AI tools are illustrated in Figure 13.2.

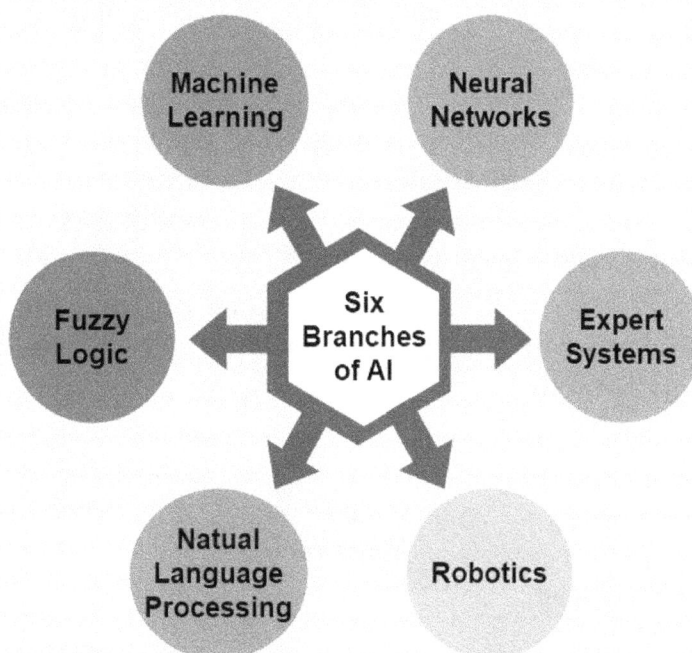

Figure 13.2 Branches of artificial intelligence.

Each AI tool has its own advantages. Using a combination of these models, rather than a single model, is recommended. AI systems are designed to make decisions using real-time data. They have the ability to learn and adapt as they make decisions. Chatbots, as part of AI devices, are NLP systems acting as a virtual conversational agent.

13.3 WHAT IS A CHATBOT?

A chatbot is a computer program that imitates human conversation. In other words, a chatbot is an automated program that interacts with customers like a human would.

The term "chatbot" comes from "chatterbot," a name coined by inventor Michael Mauldin in 1994. Chatbot is variously known as chatterbot, AI chatbot, AI assistant, intelligent virtual assistant, virtual customer assistant, digital assistant, conversational agent, conversational AI, virtual agent, conversational interface, etc. Chatbots are built and used for different purposes. AI chatbot builders often use Java, C++ or Python.

How does a chatbot work? At the heart of a chatbot lies natural language processing (NLP), a branch of AI. Chatbots process the text presented to them by the user, responds based on a complex series of algorithms that interpret and identify what the user said, infers what they mean and/or want, and determine appropriate responses. The brain of a chatbot or database consists of an organized list of suitable replies given for possible questions that may come up from user's end. Thus, an AI chatbot is fed input data, which it interprets and translates into a relevant output. The input could be voice, text, gesture or touch. One can ask questions using voice or text in the same way one would ask a person and the chatbot will respond in a conversational manner. Designed to simulate the way a human would behave as a conversational partner, chatbot systems typically require continuous tuning and testing. An AI chatbot is usually trained to operate more or less on its own. It has a key advantage of being able to learn a lot about its users. A chatbot learns from every conversation it has with the users or customers [6]. Figure 13.3 shows how chatbot works.

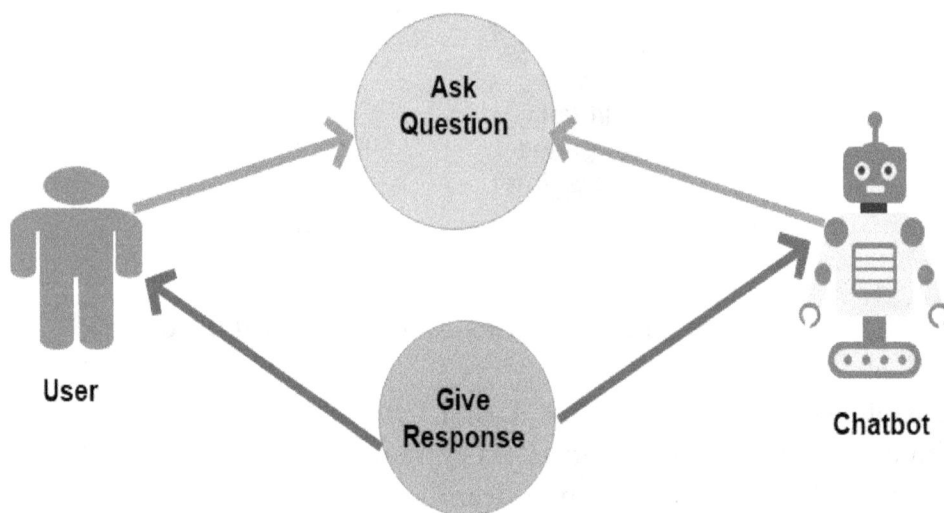

Figure 13.3   How chatbot works.

Today, you do not need to be a seasoned programmer in Java, C++ or Python to develop your own AI-based chatbot. You can make your very own chatbot in **Facebook Messenger** or other platforms such as Skype, Slack, Telegram, and Kik. Facebook Messenger allows developers to place chatbots on their platform. People unfamiliar with coding can create a chatbot using simple drag-and-drop. Once the chatbot is developed, it is crucial to test the chatbot for its intended purpose.

There are basically two types of chatbots in use today:

1. **Rule-based Chatbots:** These tend to be simpler systems that use predefined commands/rules to answer queries. If the user asks anything outside of those commands, the bot just does not know and cannot answer correctly. Currently, the rule-based chatbots are very popular for customer service since they are easy to build.

2. **AI-based Chatbots:** These use NLP and ML algorithms to improve their functionality and become smarter over time. AI-powered chatbots are usually more complex than rule-based chatbots and tend to be more conversational. They incorporate AI, the ability to understand language, not just commands, and the capacity to learn. AI chatbots are variably known as bot, talkbot, bot, intelligent chatbot, conversation bot, interactive agent, artificial conversation entity, or virtual talk chatbot [7]. Figure 13.4 shows a chatbot based on three key structures in AI.

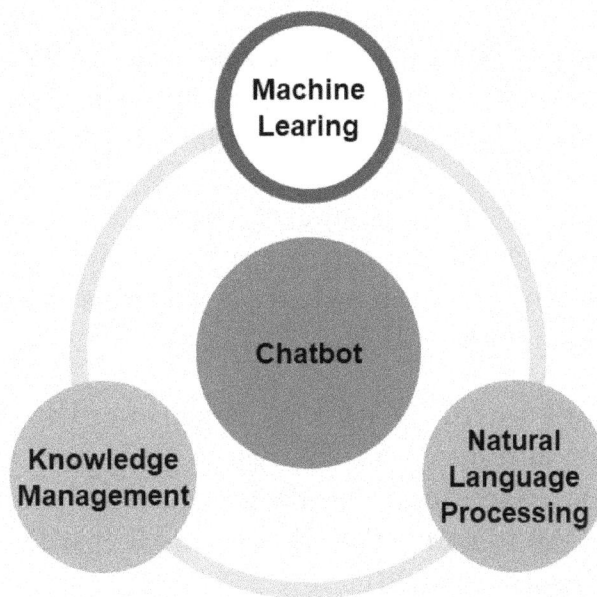

Figure 13.4   A chatbot based on three key structures in AI.

Both types of chatbots have their advantages and disadvantages.

In essence, an AI chatbot is fed input data, the message sent by the user. The message is then processed by NLP and chatbot interprets and translates it into a relevant output according to the existing database, creating a forth-and-back conversation. Artificial Intelligence gives a human touch to every conversation chatbot makes. AI uses two important tools that help the chatbot to understand, react, and learn from every interaction: machine learning (ML) and natural language processing (NLP). Machine learning helps the chatbot to learn from every conversation it has with the users. Through the use of NLP, chatbots are able to respond to a remarkable variety of customer inquiries with correct information. A chatbot using NLP allows users to pose a question as they would to another human being. AI-enabled chatbot makes the bot capable and intelligent to answer complex queries.

13.4 SOME NOTABLE CHATBOTS

AI chatbot uses ML and NLP` to deliver near human like conversational experience. Chatbots have been around since the 1960's. Amazon, Apple, Google, Microsoft, and Slack support chatbots. The following is a handpicked list of chatbots [8-11]:

- **ELIZA** was introduced by Joseph Weizenbaum in 1966 at MIT AI Laboratory. It was designed to see if chatbot systems could fool users that they were real humans. Since then, chatbots have gone through a lot of improvement.

- **ALICE** is an acronym which stands for Artificial Linguistic Internet Computer Entity. It is one of the first bots to go online. It was developed by Dr. Richard Wallace in 1995. None of today's chatbots would have been possible without the groundbreaking work on ALICE.

- **Watson Assistant** (or IBM Watson) is named after IBM's first CEO, Thomas, J. Watson. It was introduced in February 2017, and it allows you to build conversational interfaces into any device, channel, use, or any cloud. It was originally developed to compete on program, "Jeopardy!", where it defeated two of the former champions in 2011.

- **Siri** first came to the public's attention in February 2010 when it was launched as a new iPhone app.

- **Alexa** was launched by Amazon in 2015 as mobile voice assistant.

- **Google Now** was developed by Google, created specifically for the Google Search Mobile App.

- **Cortana** is an intelligent personal assistant that was developed by Microsoft in 2015.

- **Gengobot** is a chatbot-based dictionary application about multi-language grammar developed in Japan.

- **Roof Ai** is a chatbot that helps real-estate marketers to automate interacting with potential leads. It is an accurate bot that realtors would find indispensable.

- **Ada** is an AI-powered chatbot that makes easy for your team to solve customer service inquiries quickly.

Figure 13.5 shows a brief history of chatbots.

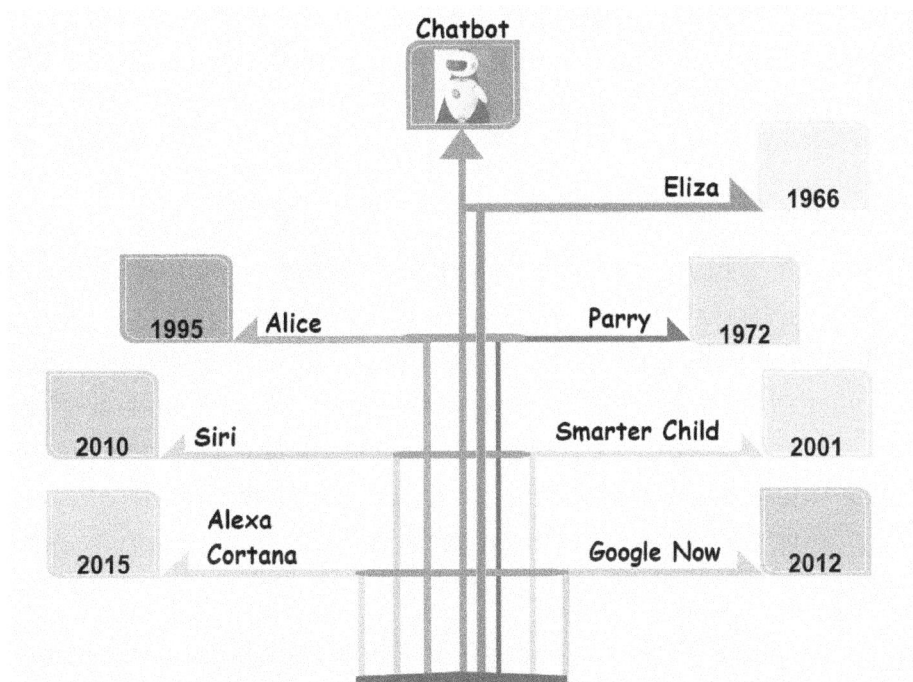

Figure 13.5   A brief history of chatbots.

13.5 APPLICATIONS OF AI IN CHATBOTS

Chatbots generally are used for the intelligent assistant applications. They can act as automated conversational agents, capable of promoting health, providing education, and potentially prompting behavior change. Chatbot applications streamline interactions between humans and services. Chatbots are used in dialog systems for various purposes including customer service, request routing, or information gathering. They can be classified into usage categories that include commerce, education, entertainment, finance, health, news, and productivity. Some common applications are explained as follows [12]:

- **Business***:* The business community has capitalized on chatbots, and they are increasingly employed in businesses. There are banking bots, customer service bots, bots for taking purchase orders, etc. Many banks, insurers, media companies, ecommerce companies, airlines, hotel chains, retailers, healthcare providers, government entities, and restaurant chains have used chatbots to answer simple questions, increase customer engagement, promote their products and services, offer different ways to order from them, and provide faster and cheaper assistance to their customers. In the recruiting and human resources departments, chatbots are being used to pre-screen, engage, and acquire candidates. Chatbots have brought a revolution in the business communication process as well as helped in attaining customer satisfaction. The major benefit of using chatbots in banking include cost reduction, financial advice, and 24/7 support and availability on your website.

- **Customer Service:** Today, AI chatbots are becoming an essential piece of the workplace and are commonly being used for customer service. They are automated programs that interact with customers and cost next to nothing to engage with. They help add convenience for customers. AI chatbots allow you to understand the frequent issues your customers have. They save your time, money, and give better customer satisfaction. They learn from customer interactions and can pick up patterns in user behavior. The future of customer service is dictated by the ability of the smart chatbots to effectively understand users' requirements and deliver appropriate responses.

- **Healthcare:** Chatbots are appearing in the healthcare industry and are increasing been used. Chatbots are beneficial for scheduling doctor appointments, locating health clinics, or providing medication information. A chatbot aims to make medical diagnoses faster, easier, and more transparent for both patients and physicians. Chatbots can increase access to healthcare, improve doctor–patient and clinic–patient communication, manage remote testing, medication adherence monitoring or teleconsultations.

MedWhat is a chatbot that aims at making medical diagnoses faster, easier, and more transparent for both patients and physicians – think of it like an intelligent version of WebMD to which you can talk. Future chatbots will reliably, accurately, and cheaply diagnose patients, recommend treatments, and even prescribe medication for the patient [13,14]. Figure 13.6 illustrates some applications of chatbots in healthcare.

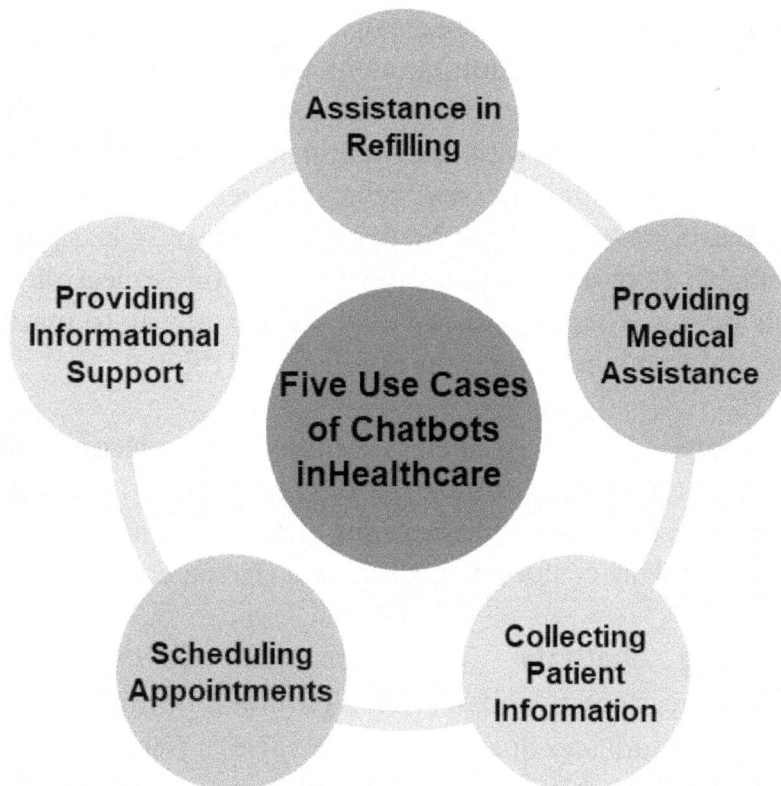

Figure 13.6 Five applications of chatbots in healthcare.

• **Government:** Motivated by the efficiency gains reported by private organizations, chatbot technology has also started being adopted by government agencies. As the public sector is seeking to improve citizens' services and government functions, more sophisticated chatbot applications have emerged, mainly targeted to automated information provision by governments [15].

• **Library:** Many libraries continue to see technology as a way to make up for reductions in funding. Chatbots offer a self-service option for our online customers in the context of information services.

Chatbots can also ease the burden of basic or routine questions so that library staff can focus their attention on more demanding inquiries and duties. Also, 24/7 availability allows users to immediately access library services at their convenience, even when the library is closed [16].

- **Education:** The educational sector can greatly benefit from chatbot technology. Chatbot has a long history of use as pedagogical agents in educational settings. It can improve productivity, communication, student learning experience, efficient teaching assistance, and minimize ambiguity from interaction. Building chatbots in education (Edu Chatbots) that can interact with learners through natural language has great value and has attracted the attention of researchers in recent years. Education chatbots improve communication, increase productivity, and simplify learning interaction [17]. Chatbots are available 24 hours 7 days a week for someone to learn foreign language on their own mobile device. They provide an easy and cheap way to master the basics of a foreign language and gain some communication skills [18].

- **Communications:** The key source of revenue for the telecommunications industry requires regular contact with customers to discuss product availability. Chatbots or intelligent virtual assistants are transforming the telecommunications industry by supporting millions of daily customer requests and transactions.

- **Troubleshooting chatbots**: These are chatbots that aim at automating troubleshooting contact centers and services. Troubleshooting chatbots have inspired considerable interest in industry due to their cost savings potential.

Other applications of AI chatbots include ecommerce, online banking, hospitality industry, restaurant, customer care, law. politics, toys, troubleshooting, and social media.

13.6 BENEFITS

A chatbot is an AI creature or pseudo-creature which can converse with humans. Its objective is to stay connected, simulate conversation, and minimize human intervention. Chatbots can efficiently conduct a dialogue and usually replace other communication tools such as email, or phone. Chatbots can help you order a pizza, make hotel reservations, schedule doctor appointments, or guide you through a complex sales process.

AI chatbot can act as automated conversational agents, capable of promoting health, providing education, and potentially prompting behavior change.

Other benefits of chatbots include [16]:

- They can understand a spoken language and use speech communication as user interface.

- They make asking questions easier (by providing a natural language interface)

- They provide instant responses.

- They can personalize the user experience

- They are anonymous (which encouraged shy users or those who thought their questions might be "stupid")

- They provide a marketing tool for reference services

- They are patient and polite and remain unruffled by rude customers, high traffic, or repeated requests

- They save time, money, and give better customer satisfaction

- They can deliver near human-like conversational experience

- They increase customer satisfaction

- They support customization without writing any code

- They satisfy the human need for sociability.

13.7 CHALLENGES

In spite of the benefits of AI chatbots, they have some important limitations. There are privacy and security concerns, lack of personality and lack of research resulting in errors and financial expenses. Besides the risk of implementing chatbots, there are high financial costs associated with acquiring, updating, and hiring specialists [18]. Human–chatbot communication has noticeable differences in quality in comparison to the human–human discussion. Other challenges include [19].

- As the database, used for output generation, is fixed and limited, chatbots can fail while dealing with an unsaved query.
- A chatbot's efficiency highly depends on language processing and is limited because of irregularities, such as accents and mistakes.
- Chatbots are unable to deal with multiple questions at the same time and so conversation opportunities are limited.
- Chatbots require a large amount of conversational data to train.
- Chatbots have difficulty managing non-linear conversations that must go back and forth on a topic with a user.

- As it happens usually with technology-led changes in existing services, some consumers are uncomfortable with chatbots due to their limited understanding, making it obvious that their requests are being dealt with by machines.

- Some see chatbots as malicious by design and a potentially criminal dimension to these agents.

- Some hate being addressed by machines and the fact that chatbot or computer voices are mostly female.

It is needless to say that these chatbot limitations can frustrate customers.

13.8 FUTURE OF AI CHATBOTS

Chatbots are becoming an indispensable part of our digital life. They are becoming a ubiquitous trend in several sectors such as medicine, business, ecommerce, entertainment, industry, and education. For AI and chatbot, the future is coming one way or another, and that cannot be avoided. Although it is hard to predict the future of chatbots, they will be part of the digital transformation. The success of AI chatbots in the future depends on their intelligence, knowing the users and improving based on their interaction with users [20].

There are many types of chatbots available depending on the complexity and they can be classified as follows [11]:

- **Traditional chatbot:** Traditional chatbots are driven by automation.

- **Current chatbot:** Current chatbots are driven by back-and-forth communication between the system and humans. They have the ability to maintain both system and task contexts.

- **Future chatbot:** Future chatbots can communicate at multiple levels with automation at the system level. They have the ability to maintain the system, task, and people contexts. There is a possibility of introduction of master bots and eventually a bot OS.

Figure 13.7 shows different areas where chatbots are predicted to be used in future [21].

Figure 13.7   What one can predict to use a chatbot for [20].

One should also expect to see enterprises planning for an intranet of conversational AI applications.

13.9 CONCLUSION

Chatbots are basically software applications that mimic written or spoken human speech for the purposes of simulating a conversation or interaction with a real person.

Empowered by AI, chatbots are emerging as new technologies with great business potentials. The last few years has witnessed an exponential growth of tools to design, build, deploy, manage, and monetize chatbots. In recent years, the use of chatbots has evolved in numerous fields such as business, ecommerce, education, healthcare, and entertainment.

Chatbots will handle the increase in customer inquiries, while deploying a remote workforce. Chatbot development has increased while in many cases its purpose remains loosely defined.   Chatbots are one of the ongoing trends in the global market. Companies adopt Chatbots in order to offer better services to their customers. Chatbots.org provides more than 1,350 chatbots and virtual agents in use around the world. Adopting a chatbot is simple and straightforward with the various technologies and tools available today.

Although AI chatbot technology is still in its developmental phase, AI chatbots are getting smarter over time. Today, organizations are becoming more comfortable with integrating chatbots into their activities. Whether you like them or hate them, chatbots are here to stay. They are becoming more human. More information on AI chatbots can be found in the books in [22-26] and in related periodicals: Chatbots Magazine and Machine Learning with Applications.

REFERENCES

[1] **M. C. Palmer**, "Ethical considerations of legal chatbots," https://www.attorneyatwork.com/legal-chatbots-ethical-considerations-for-law-firms/

[2] **O. Zahour et al.**, "A system for educational and vocational guidance in Morocco: Chatbot E-Orientation," Procedia Computer Science, vol. 175, 2020, pp. 554–559 .

[3] **S. Greengard**, "What is artificial intelligence?" May 2019, https://www.datamation.com/artificial-intelligence/what-is-artificial-intelligence.html

[4] https://in.pinterest.com/pin/828662400161409072/

[5] **N. Tyagi**, "6 Major branches of artificial intelligence (AI)," January 2021, https://www.analyticssteps.com/blogs/6-major-branches-artificial-intelligence-ai

[6] **K. M. Kaka-Khan**, "Building Kurdish chatbot using free open-source platforms," UHD Journal of Science and Technology, vol. 1, no. 2, August 2017, pp. 46-50.

[7] **O. O. Oyebode** and **R. Orji**, "Likita: A medical chatbot to improve healthcare delivery in Africa," https://www.researchgate.net/publication/330522151_Likita_A_Medical_Chatbot_To_Improve_HealthCare_Delivery_In_Africa/link/5c45f692458515a4c736683d/download

[8] **C. Chi**, "12 of the best AI chatbots for 2021," https://blog.hubspot.com/marketing/best-ai-chatbot

[9] "Chatbots: The definitive guide (2020)," https://in.pinterest.com/pin/230879918384639777/

[10] **D. Shewan**, "10 of the most innovative chatbots on the web," January 2021, https://www.wordstream.com/blog/ws/2017/10/04/chatbots

[11] Great Learning Team, "Basics of building an artificial intelligence chatbot,' May 2020, https://www.mygreatlearning.com/blog/basics-of-building-an-artificial-intelligence-chatbot/

[12] "Chatbot," Wikipedia, the free encyclopedia https://en.wikipedia.org/wiki/Chatbot

[13] **M. N. O. Sadiku, P. O. Adebo, A. Ajayi-Majebi**, and **S.M. Musa**," Chatbots in healthcare," International Journal of Trend in Research and Development, vol. 7, no. 3, May-June 2020, pp. 91-93.

[14] **N. Babrovich**," How medical chatbots can be used in the healthcare industry," https://www.scnsoft.com/blog/chatbots-in-healthcare

[15] **A. Androutsopoulou et al.**, "Transforming the communication between citizens and government through AI-guided chatbots," Government Information Quarterly, vol. 36, no. 2, April 2019, pp. 358-367.

[16] **M. L. McNeal** and **D. Newyear**, "Chapter 1: Introducing chatbots in libraries," Library Technology Reports, vol. 49, no. 8, November-December 2013. Also available online: https://journals.ala.org/index.php/ltr/article/view/4504/5281

[17] **P. Smutny** and **P. Schreiberova**, "Chatbots for learning: A review of educational chatbots for the Facebook Messenger," Computers & Education, vol. 151, July 2020.

[18] **I. Dokukina** and **J. Gumanova**, "The rise of chatbots – New personal assistants in foreign language learning," Procedia Computer Science, vol. 169, 2020, pp. 542–546.

[19] **E. Carter** and **C. Knol**, "Chatbots — An organisation's friend or foe?" Research in Hospitality Management, vol. 9, no. 2, 2019, pp. 113-116.

[20] "Power through AI and automation with chatbots,"
\ https://www.infosys.com/services/microsoft-dynamics/documents/ai-automation-chatbots-web.pdf

[21] **S. Patel**, "Top 12 chatbots trends and statistics to follow in 2021," January 2021, https://www.revechat.com/blog/chatbots-trends-stats/

[22] **S. Janarthanam**, Hands-on Chatbots and Conversation UI Development. Birminghan, UK: Packet Publishing, 2017.

[23] **O. Muldowney**, Chatbots After 2020: All You Need to Know About AI, NLP and Chatbots in The New Era. Independently Published, 2020.

[24] **O. Muldowney**, Chatbots: An Introduction And Easy Guide To Making Your Own. Curses & Magic, 2017.

[25] **S. Raj**, Building Chatbots With Python: Using Natural Language Processing and Machine Learning. Apress, 2018.

[26] **L. Safko**, The Artificial Intelligence Chatbot: Unexpected Positive Consequences. Innovative Thinking, 2019.

CHAPTER 14

AI IN SOCIAL MEDIA

"Social media is the ultimate equalizer. It gives a voice and a platform to anyone willing to engage." *Amy Jo Martin*

14.1 INTRODUCTION

Artificial intelligence (AI) is the cognitive science that deals with intelligent machines which are able to perform tasks heretofore only performed by human beings. It is a branch of computer science that deals with the capability of a machine to imitate human intelligent behavior. It has the potential to help tackle some of the world's most challenging social problems. It is mainly concerned with applying computers to tasks that require knowledge, perception, reasoning, understanding, and cognitive abilities [1]. AI tools can be trained to leverage individual behaviors, preferences, beliefs, and interests to personalize experiences. They can teach machines to be like humans. They can provide them the ability to see, hear, speak, move, and write. AI can learn these habits at a rate much faster than humans. AI tools are being used across industries to automate and improve the efficacy of diverse activities.

Social media have become an indispensable part of life. Consumers constantly interact with social media. Modern social media, also known as social networking, include Facebook, Twitter, Instagram, Pinterest, and YouTube [2]. AI is a fundamental component of how today's social network's function. The use of AI in social media is growing in an unprecedented way and is constantly transforming social media. Social media is one of the major sectors where marketers can both skyrocket performance and efficiencies by using AI. With the help of AI, data about your activity on social media is continuously being compiled and analyzed. Social media, in combination with big-data analysis tools, is currently being used to infer social behavior and derive tendencies [3].

This chapter explores the impact of various artificial intelligence tools on social media companies. It beings by briefly reviewing AI. It presents an overview of social media and popular social media platforms. It covers how AI is transforming social media and some applications of AI in social media. It highlights the benefits and challenges of AI in social media. The last section concludes with comments.

14.2 REVIEW ON ARTIFICIAL INTELLIGENCE

Artificial intelligence (AI) refers to computer systems that mimic human cognitive functions. It is a field of computer science that deals with intelligent machines. The term "artificial intelligence" was first used at a Dartmouth College conference in 1956. The main goal of AI is to enable machines to perform complex tasks that typically require human intelligence [4]. In simple terms, AI attempts to clone human behavior. An important feature of AI technology is that is can be added to existing technologies. AI is now one of the most important global issues of the 21st century. It is poised to disrupt our world and change processes and developments in the field of science, engineering, education, business, and agriculture.

The concept of AI is an umbrella term that encompasses many different technologies. AI is not a single technology but a collection of techniques that enables computer systems to perform tasks that would otherwise require human intelligence [5]. The major disciplines in AI include

- Expert systems
- Fuzzy logic
- Neural networks
- Machine learning (ML)
- Deep learning
- Natural Language Processors (NLP)
- Robots

These AI tools are illustrated in Figure 14.1.

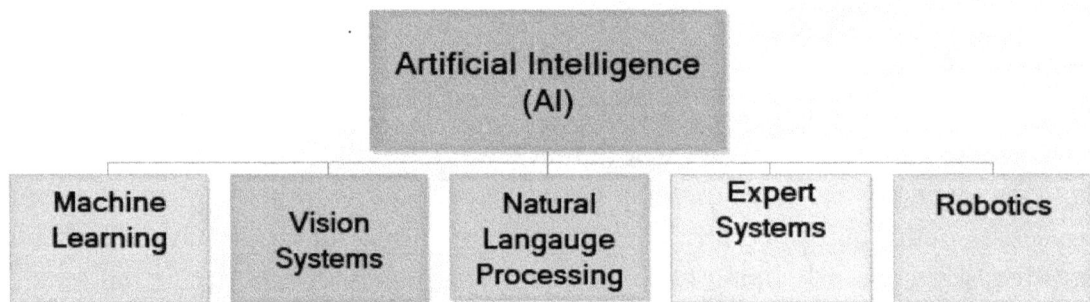

Figure 14.1 Branches of Artificial intelligence.

Each AI tool has its own advantages. Using a combination of these models, rather than a single model, is recommended. AI systems are designed to make decisions using real-time data. They have the ability to learn and adapt as they make decisions.

14.3 SOCIAL MEDIA BASICS

Traditional social media include written press, TV, and radio. Modern social media, also known as social networking, include Facebook (Facebook, Inc, Menlo Park, California, USA), Twitter (Twitter Inc, San Francisco, California, USA), YouTube (San Mateo, California, USA), LinkedIn (Sunnyvale, California, USA), Instagram (Facebook, Inc, Menlo Park, California, USA), and Pinterest (San Francisco, California, USA). Both the traditional and modern social media are illustrated in Figure 14.2.

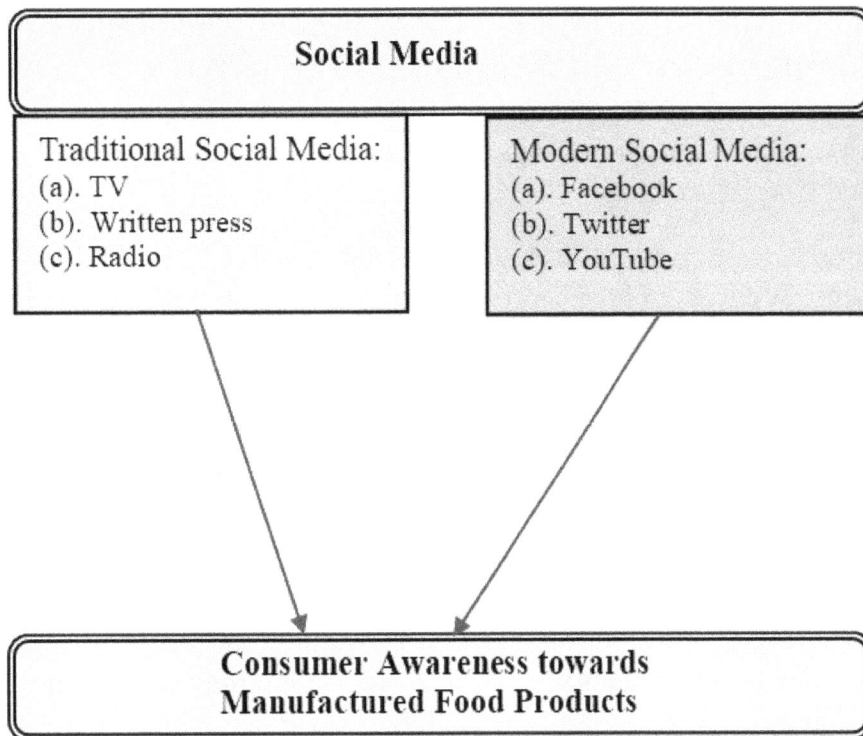

Figure 14.2   Traditional and modern social media.

Modern social media began in 1978 by Ward Christensen and Randy Suess who created bulletin board to inform friends of meetings, announcements, and share information. Since then, social media has become an integral part of our life. Social media gives companies another means of reaching people in ways that traditional media cannot. They allow your company to boost their brand. Companies are better leveraging social media through AI. Companies who fail to invest in having a strong presence on social media will soon realize they missed out on a serious competitive advantage [6,7].

Social media (also called Web 2.0 or social networking) refers to Internet-based and mobile-based tools that allow individuals to communicate, share ideas, send personal messages, and images.

Web 1.0 is the "read-only Web." Consumers are allowed to read information created by the provider of the online information. Web 2.0 allows users to create information, send posts and share audio, graphics, text, and video online [8].

Social media are computer-mediated communication tools that enable users to share and consume content through varied modalities such as text, image, and video [9]. Recently, the use of social media has been extended to the healthcare field. Healthcare professionals now use SM as part of their daily activities. Social networking sites allow users to share ideas, activities, events, and interests. The majority of those who use social networking sites use them to form self-aggregated interest groups for fundraising, awareness, marketing, and general support.

SM sites may include wikis, blogs, and social networks.

Wikis: These are easy-to-publish websites. They can be quickly and easily edited by multiple users. Wiki" is a Hawaiian term meaning "quick." Wikipedia happens to be the most commonly used wiki in the medical community as healthcare professionals use Wikipedia to find medical information. However, Wikipedia sometimes contains factual errors that lacks depth compared to traditionally edited, peer-reviewed information sources.

Blogs: These are the oldest, most established, and evaluated form of social media. They provide the opportunity to publish large amounts of information in a variety of media (text, video, and audio) in an open forum. Blogs have been used by healthcare workers for peer-to-peer communication. Medical blogs generally target one of two different audiences: patients or providers. Microblogs provide dynamic and concise form of information exchange through social media.

Social Media: Social media platforms such as Facebook allow individuals to post photos and messages and share them with friends, relatives, and acquaintances all over the world. Media sharing sites comprise social media tools that are optimized for viewing and sharing. They are great resources for education, community building, marketing, and research. They have become encyclopedic resources.

Today, many social media tools, including social networking sites, blogs, microblogs, wikis, media-sharing sites, are influential in our everyday life and are available for health care professionals (HCPs).

Mobile social media (MSM) has emerged as the combination of social networking and mobile technologies. It is becoming a global phenomenon as it enables IP-connectivity for people on the move. It is mediated by mobile devices such as smartphones, tablets, or laptop computers. It refers to social media applications such as Facebook, LinkedIn, Instagram, Myspace, and Twitter that are delivered via mobile devices. These popular social media platforms have made mobile apps for their users to have instant access from anywhere at any time [10].

14.4 POPULAR SOCIAL MEDIA

Social media sites serve different purposes which may include blogs, social networks, video- and photo-sharing sites, wikis, or a myriad of other media. These uses can be grouped by purpose as [11].

- Social networking (Facebook, Instagram, Twitter, Snapchat)
- Professional networking (LinkedIn)
- Media sharing (YouTube, Flickr)
- Content production (blogs [Tumblr, Blogger] and microblogs [Twitter, Instagram])
- Knowledge/information aggregation (Wikipedia)
- Virtual reality and gaming environments (Second Life)

Social media is consumer-generated media that covers a variety of new sources of online information, created, and used by consumers with the intent on sharing information with others. It employs mobile and web-based technologies to create, share, discuss, and modify consumer-generated content. Consumers are most likely to leverage their power in social media to be more demanding of marketers [12]. Social networks have become an important healthcare resource. The most popular social media platforms are described here [13,14].

Facebook: This is the most popular social media in the US and the rest of the world. It has the largest user base of any social media platform, with 2 billion active monthly users. It was launched in February 2004 by Mark Zuckerberg as a Harvard social networking site, expanding to other universities and eventually to everyone. Facebook can sensitize individuals (consumers) about many products and services. A company can use Facebook to communicate their core values to a wide range of customers. Marketing strategists have found Facebook to be useful because it covers a range of personal and organizational interests. Facebook groups can use social media for healthcare professionals and patients to interact. Facebook Advertising costs extra, and this is how Facebook makes money.

Twitter: Twitter was launched in July 2006 to provide a microblogging service. Twitter provides a real-time, Web-based service which enables users to post brief messages for other users and to comment on other user posts. Tweets are extracted from Twitter. A tweet is a small message of no more than 140 characters that users create in order to communicate thoughts. Microblogging is a newer blog option made popular by Twitter. Many Twitter posts (or "tweets") focus on the minutiae of everyday life. Twitter has been used at medical conferences to enhance speaker presentations by posting real-time comments and feedback from the audience.

LinkedIn: This is a networking website for the business community. It is a professional network that provides a platform for professionals to participate in networking with each other. By setting up an account on LinkedIn one can link with professional individuals of similar interests. LinkedIn remains the most popular social networking site for organizations to recruit new employees. It allows people to create professional profiles, post resumes, and communicate with other professionals. LinkedIn is where companies see their largest audiences. Many regards LinkedIn as a strictly professional networking site and would never post personal information there.

YouTube: YouTube has established itself as social media. It was launched in May 2005. This is a video sharing platform where many people can discover, watch, and share user-generated videos. It is a website of participatory culture. It has become the most successful Internet website providing a short video sharing service since its establishment in early 2005. Since YouTube is a Google property, its required to have a Google account to prior to signing up for a YouTube account. YouTube may serve as home to aspiring filmmakers who might not have industry connections. YouTube can be both a blessing and a curse for some companies.

MySpace: This social networking site bases its existence on advertisers who are paying for page views. It has a lot that users could do. There are MySpace sites in United Kingdom, Ireland, and Australia.

Instagram: This is an image-based social media platform with more than 700 million active monthly users. The design is centered on a visual mobile experience. Instagram allows a simple and creative way to capture, edit, and share photos, videos and messages with friends and family.

Snapchat: This leverages an AI technology in the form of computer vision to monitor your facial features and then superimpose filters for your face in real-time.

Pinterest: The major reason many users love Pinterest is because of the personalized content that it shows. Pinterest Lens allows users to take photos and use them to search for related items rather than entering keywords. Over 80% of Pinterest's active users make purchases via the platform because of the hyper-personalized content that Pinterest offers.

WeChat: This is a Chinese social medium known as Weixin in Chinese. It was launched in 2011 by Tencent, the Chinese largest Internet company. It offers instant messaging service for smartphones and allows exchange of videos among mobile phone users. Individuals, governments, and organizations freely apply for WeChat accounts [15].

Other social media include Reddit, Flickr, WeChat, and Vine Camera.

14.5 HOW AI IS TRANSFORMING SOCIAL MEDIA

AI is a key component of the popular social networks you use every day. In the digital age, AI is constantly transforming social media. It can handle certain types of social media creation and management in minutes. AI acts as a way for brands to scope the vast pool that is social media. Various tech giants have branded their marketing AI in their own image. For example, IBM has Watson, while Salesforce has Einstein. Figure 14.3 shows the impact of AI in social media [16].

Figure 14.3   The impact of AI in social media [16].

Here are some examples of how social media platforms are leveraging AI [17]:

Facebook does almost everything on its platform like proposing your content, recognizing your face, suggesting friends, and targeting ads to users through advanced machine learning. The platform uses machine learning and AI for serving you the content of your interest. Facebook uses a variety of AI tools to heighten each user's experience.

Instagram is a unique platform with many facets. It identifies and suggests visuals and images using artificial intelligence. The first way that Instagram uses artificial intelligence is seen on its Explore page. This company aims to make its users feel like a part of a worldwide community. Instagram uses AI to filter spam. This system currently works with nine languages. The analysis of Instagram offers numerous benefits for marketing. AI enables Instagram to process content generated by its millions of users, which is manually impossible. It is also making selling on Instagram easy. Instagram uses AI to identify different visuals.

Twitter is working on using AI and machine learning to categorize and rank every single tweet. It uses AI to detect a face from a complete image. It could translate into shift in the way that people currently view tweets within the chronological timeline format. It also uses AI to process copious amounts of data and root out fraud, propaganda, and hateful comments.

LinkedIn relies on machine learning and AI to predict suitable candidates for a particular job role. It uses AI technology to parse massive data produced by social media audiences and recommend jobs.

Pinterest is a huge community of users, which implies that the platform is doing a great job. Pinterest has evolved from a mere bookmarking platform to a huge ecommerce arena. AI for Pinterest works on a deep learning algorithm, which is based on neural networking. This means every image on the platform is attached to a neural network based on a theme. It is now serving up personalized recommendation.

YouTube is owned by Google and uses AI to remove violent and illegal videos. Data shows that 76% of videos (or objectionable contents) were automatically identified and flagged by AI

These are just a few examples of how social media is using AI. It is evident that AI has become a fundamental part of social media.

14.6 APPLICATIONS OF AI IN SOCIAL MEDIA

Social media has moved away from its traditional role of being a platform where humans interact and connect with each other. Today, smart companies are using social media for ecommerce, customer service, marketing, public relations, and more. The applications of AI in social media companies are many. Examples of how AI is used on social media platforms include analyzing text, analyzing pictures, detecting spam, social insights, advertising, and data gathering. Some of these applications are discussed as follows [17-19].

- **Social Media Advertising:** AI is almost at the helm of ruling the realm of marketing technology. There are AI tools that will write social media ads for you. Many social media platforms have built-in advertising systems that one can use to enhance the results of their marketing operations. While social media platform gives individuals and businesses to connect to people, it also allows brands to run paid advertisements based on behavioral targeting and demographics. AI can write Facebook and Instagram short-form ad for you.

- **Social Media Marketing:** The success of social media marketing largely depends on influencers, who can promote any products to their loyal followers. But finding the right influencers can be an uphill task. Artificial intelligence can help marketers track down and choose the most profitable influencers based on follower count, personal branding, and relevant content. AI is transforming the social media marketing landscape. It will enable you to stand out from the competition, forge a stronger connection with your customers, and improve your bottom line. Marketers have a range of tools at their disposal for understanding customers and prospects on social media. The combination of social media marketing with AI is popularly known as "social artificial intelligence." Many social networks give marketers an unprecedented ability to run paid ads to platform users. With the assistance of AI, marketers are able to identify customers, curate content, and advertise better. AI tools can write Facebook and Instagram ads for you. Marketers should enhance the digital experience of buyers but not at the expense of their convenience and privacy. Human touch still remains the most desirable aspect in social media marketing. The next-generation machine learning will take branding and marketing activities at a higher level. Figure 14.4 illustrates AI in social media marketing [20].

Figure 14.4 AI in social media marketing [20].

- **Social Media Management:** One of the most frequent uses of AI is on social media management. Today, most businesses have a presence on multiple social media networks. This complicates the social media management process.

- **Social Insights:** AI-powered tools can deliver insights from your brand's social media content, profiles, and audience. AI-powered social media intelligence can help companies improve brand equity, detect consumer trends, understand target audiences, and analyze your posts and the posts of other companies to recommend what to post. One can use the unparalleled insights that AI technologies provide to increase productivity, identify new trends, reach a wider audience, figure out what works for your niche, track performance, and optimize campaigns in real-time.

- **Security and Justice:** This domain involves preventing crime and other physical dangers, as well as tracking criminals and mitigating bias in police forces. It focuses on security, policing, and criminal-justice issues. AI can be used to identify tax fraud using alternative data such as browsing data, retail data, or payments history.

- **Automation:** By definition, AI includes elements of intelligent automation. Automation capability is one of the biggest advantages of using AI in social media. Automation will make your business far more productive. Examples of tasks to automate include social listening, social engagement, content scheduling, content republishing, and analytics tracking.

- **Social Listening:** This is another area in which AI is making great impact. Social listening is a technique that is often used to monitor a business' social media channels and help listen to what people are saying about your products and services. Through social media listening tools, businesses can efficiently analyze thousands of conversations and identify patterns. They can use the information for their overall marketing strategy.

- **Chatbots:** Chatbots have benefited sponsors on social media in many ways. AI-powered chatbots are benefiting digital marketers in many ways. Companies running over social media can use AI-powered chatbots to answer their customers' queries in no time. AI tools help companies to automatically reply to messages by developing user-interactive AI chatbots. Brands can also use chatbots for businesses to provide personalized support to shoppers. With this, businesses can improve customer experience to a significant level. Figure 14.5 depicts chatbots as AI in digital marketing [21].

Figure 14.5 Chatbots as AI in digital marketing [21].

- **Social Media Robots:** AI is a way of making a computer-controlled robot, or a software think intelligently, in a similar manner an intelligent human brain thinks. It is based on the study of how a human brain thinks, and how humans learn, decide, and work while trying to solve a problem.

Other areas of applications of AI in social media include customer behavior, customer analysis, consumer engagement, social media monitoring, competitive analysis, content creation, content curation, sentiment analysis, social media analytics, image recognition, facial recognition, and text understanding.

14.7 BENEFITS

Artificial intelligence in social media enhances the user experience and overall effectiveness of social platforms. It has the potential to transform how brands create and manage social media marketing. Figure 14.6 illustrates some benefits of using artificial intelligence in social media.

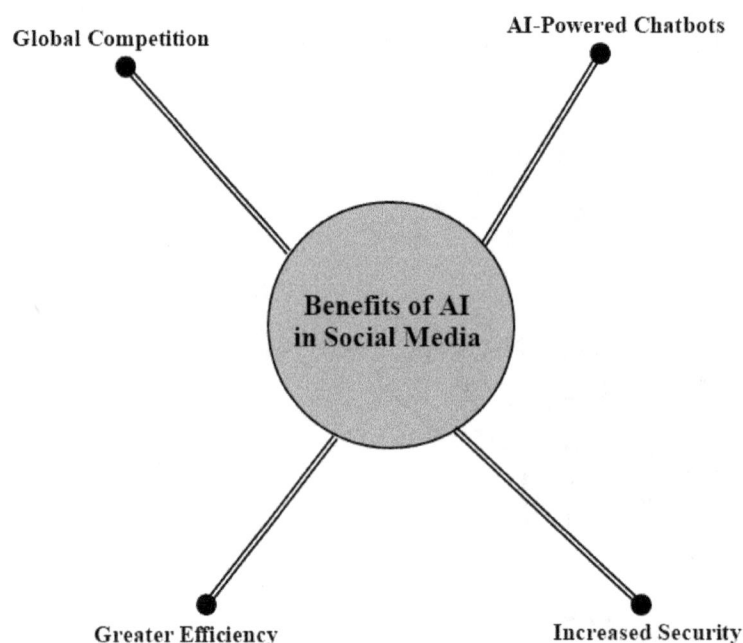

Figure 14.6 Benefits of using artificial intelligence in social media

Other benefits that AI can offer social media include the following [22,23]:

1.**Increased Audience Engagement:** AI enables businesses to get to know their audience a lot better and understand their preferences.

2. **Greater Efficiency:** By relegating cumbersome, repetitive, and mundane tasks to AI, marketers save time and energy, which can be poured into tasks that require human effort. AI is capable of increasing business productivity by 40%.

3. **Smarter Advertising:** With AI on your side, you can target prospects and customers with the right message at the right time.

4. **Refined Content Targeting:** Using AI to track your audience's behaviors and preferences on social media will give you a clear understanding of how to appeal to them.

5. **Reduced Marketing Costs With Better Return on Investment (ROI).** AI poses to be a great opportunity for social media marketers. AI can fully or partially automate time-consuming and labor-intensive tasks. It can help a social media marketer increase revenue and reduce costs.

6. **AI-powered Chatbots:** Companies running over social media can use AI-enabled chatbots to answer their customers' queries in no time. With this, businesses can improve customer experience to a significant level.

7. **Increased Security:** AI can help social media platforms to protect the user data and increase the privacy of their information. Securing the brand from attacks is important for marketers. Organizations are using AI trackers to protect their data and brand image. The AI technology can track irregular movements in the process and alerts the marketers about the risk.

8. **Cost-reduction:** Once the algorithms and action points are in place, AI can function with minimum human monitoring. This helps to reduce marketing expenditure. AI has the benefit of speed as well as reduced cost, not achievable by human effort.

9. **Incremental Revenue:** AI also assists brands to boost revenue by making processes automatic and efficient. AI provides valuable insights that aids in creating content to improve engagement.

10. **Competition-minded Tool:** AI is a competition-minded tool. In business, knowing what a competitor is up to is as important as making your own business plans. AI has been helping online marketers to track competitors' business online.

11. **Global Competition:** The global competition continues to affect revenues and profit. AI technology can be utilized in social media to maintain international competitive advantage.

12. **Advertising:** AI tools exist today that will actually write eye-catching social media ads for you. Some of the attractive features that AI can conjure before shoppers appeals

13. **Massive Data:** AI helps analyze massive data produced by social media audiences daily, to make sense of it and identify consumer behavior patterns. Businesses and marketing gurus need to make sense of the data and draw actionable insights from it to be able to leverage social media to its full potential.

14. **Global Marketing:** AI can assist your global social media marketing efforts by removing language barriers through translation.

15. **Publishing:** AI is already impacting many of the world's leading publishers. Automating content distribution on social media is just one-way publishers are using AI to augment their efficiency.

14.8 CHALLENGES

Although AI could help tackle some of the world's most challenging social problems, it is not a silver bullet. While the social impact of AI is large, certain challenges must be addressed if some of the potential is to be realized. Principles for using AI tools must be established since they can be misused. Other significant challenges are [24]:

- **Data accessibility:** Resolving this significant challenge will require a willingness, by both private- and public-sector organizations, to make data available. The data owners include telecommunications and satellite companies, social-media platforms, financial institutions, healthcare providers, and governments.

- **Shortage of talent:** There are not enough professionals or expertise to develop AI solutions. Integrating AI into social media requires high-level AI talent, people who may have PhDs or considerable experience with the technologies. Such intelligentsia are in short supply.

- **"Last-mile" implementation challenges:** This is another significant bottleneck for AI deployment for social good. Most AI models cannot perform accurately all the time, and many are described as "brittle."

- **Privacy concerns:** The risk is that financial, tax, health, and similar sensitive personal records could become accessible to illegitimate individuals. That would cause embarrassment and harm. Due to privacy concerns, AI-enabled marketing automation platforms and social media marketing processes should draw a line between being intuitive and intrusive, responsible, and unscrupulous, or trust-worthy and fraudulent.

- **Ethics:** AI may pose some ethical challenge for humanity if it reaches a very advanced stage. Amazon's AI for recruitment failed because it was biased against women.

- **Eliminating Jobs:** Some critics fear that AI will replace humans. Not really. AI is meant to collaborate and work with humans. The growth of AI will drive formation of business models and inculcate new skills in the workforce.

Whether the benefits overpower the challenges or not is up to each one of us to determine.

14.9 GLOBAL AI IN SOCIAL MEDIA

The major goal of artificial intelligence is to develop systems capable of tacking complex problems in ways that are similar to human reasoning and logic. AI and related technologies will have a transformative impact on media markets all around the world. The global AI in social media market is expanding due to factors such as high demand for better insights and increase in usage of social media for advertising. However, privacy, data security, and reliability concerns among users or businesses are major factors that are expected to hinder the artificial intelligence in social media market globally. The global artificial intelligence in social media market can be segmented based on application and geography. In terms of application, the AI in social media market can be divided into advertisement, auto traffic analysis, virtual assistant, image recognition, intelligent search, automated writing bots, and others. Based on geography, the global artificial intelligence in social media market can be categorized into North America, South America, Asia Pacific, Europe, and Middle East and Africa. The AI in social media market in North America is anticipated to expand at a substantial growth rate. Asia Pacific is expected to be a lucrative region for the global AI in social media market due to the presence of a large customer base. The global market is also characterized by the presence of several key players such as IBM, Facebook, Amazon, Microsoft, Intel, Twitter, Xero, FreshBooks, NetSuite ERP, and Oracle [25]. China increasingly uses AI in almost all sectors. Chinese social media is actually full of government criticism and discussion of sensitive topics. While China is leading, countries like Japan, Singapore, South Korea, Taiwan, and India, have also shifted their focus toward AI.

14.10 FUTURE OF AI IN SOCIAL MEDIA

Artificial Intelligence has become a household term used across popular culture, science, and technology. AI is poised to rule the entire world in the future. The impact of AI on social media has been enormous, with more than three billion social media users.

AI is helping social media platforms to manage the pool of data. It is having great impact all over the world and transforming everything. It will rule the entire world in the future. Soon, machines will be able to make the same kinds of inferences that humans can.

Artificial intelligence is constantly breaking new grounds in social media. It has a bright future in social media industry as it improves user experiences and help brands to serve them better. AI technologies will likely influence the world of social media in the following ways [26].

- AI will enable the creation of hyper-personalized marketing messages
- AI will provide more robust buyer and customer personas
- AI will make it easier to distinguish between qualified leads and unqualified leads
- AI will enable bots to converse with prospects on just about any topic

- AI technology will help social networking companies to deliver better customer experience and help marketers to target the right customers that will increase their ROI.

The best we can do is to wait and see AI as it evolves.

14.11 CONCLUSION

Artificial intelligence is an umbrella term that covers many different technologies. The impact of AI on social media is growing fast. AI is transforming how brands use social media. Today, social media platforms are thinking and acting like humans as well as rationally. Many companies already realize that AI is the way forward for progressing business. In the modern society, not having a social media account is more of an anomaly than a norm.

When it comes to AI in social media, the sky is the limit. AI will continue to influence social media networks as the technology develops and evolves. The combination of AI and social media is proving to be greatly beneficial for businesses. The future is bright and exciting for any business that is harnessing AI tools. There is no question that AI will have profound impacts on media markets. More information about the uses of AI in social media can be found in the books in [27,28] and related journals: Artificial Intelligence and AI Magazine.

REFERENCES

[1] **M. N. O. Sadiku**, "Artificial intelligence", *IEEE Potentials,* May 1989, pp. 35-39.

[2] **M. N. O. Sadiku, M. Tembely**, and **S.M. Musa**," Social media for beginners," International Journal of Advanced Research in Computer Science and Software Engineering, vol. 8, no. 3, March 2018, pp. 24-26.

[3] **H. Sarmiento**, "How artificial intelligence can benefit the social media user," May 2020,
https://medium.com/clyste/how-artificial-intelligence-can-benefit-the-social-media-user-aeaefd24e0a7

[4] **S. Greengard**, "What is artificial intelligence?" May 2019,
https://www.datamation.com/artificial-intelligence/what-is-artificial-intelligence.html

[5] **M. N. O. Sadiku, J. T. Ashaolu, A. Ajayi-Majebi, S. M. Musa**, "Artificial intelligence in social media," International Journal of Scientific Advances, vol. 2, no. 1, Jan.-Feb. 2021, pp. 15-20.

[6] **C. S. Amaravadi**, "Artificial intelligence is 340,"
https://slideplayer.com/slide/6911243/

[7] **J. A. H. Kareem et al**., "Social media and consumer awareness toward manufactured food," Cogent Business & Management, vol. 2016.

[8] **J. Sarasohn-Kahn**, "The wisdom of patients: Health care meets online social media," April 2008,
https://www.chcf.org/publication/the-wisdom-of-patients-health-care-meets-online-social-media/

[9] **M. N. O. Sadiku, N. K. Ampah, S. M. Musa**, "Social Media in Healthcare," International Journal of Trend in Scientific Research and Development, vol. 2, no. 5, June/July 2018, pp. 665-668.

[10] **M. N. O. Sadiku, P. O. Adebo,** and **S.M. Musa,**" Mobile social media," International Journal of Advanced Research in Computer Science and Software Engineering, vol. 8, no. 3, March 2018, pp. 8-10.

[11] **C. L. Ventola,** "Social media and health care professionals: Benefits, risks, and best practices," P& T, vol. 39, no. 7, July 2014, pp. 491-499.

[12] **C. Kohli, R. Surib,** and **A. Kapoor,**" Will social media kill branding?" Business Horizons, 2015, vol. 58, pp. 35-44.

[13] **G. Merchant,** "Unravelling the social network: theory and research," Learning, Media and Technology, vol. 37, no. 1, 2012, pp. 4-19.

[14] **M. N. O. Sadiku, A. A. Omotoso,** and **S. M. Musa,** "Social networking," International Journal of Trend in Scientific Research and Development, vol. 3, no. 3, Mar-Apr. 2019, pp. 126-128.

[15] **J. Xu et al.,** "Applications of mobile social media: WeChat among academic libraries in China," The Journal of Academic Librarianship, vol. 41, 2015, pp. 21-30.

[16] **A. Mufareh,** "How can artificial intelligence improve social media?" May 2020, https://www.techiexpert.com/how-can-artificial-intelligence-improve-social-media/

[17] **S. Datta,** "Social artificial intelligence: Intuitive or intrusive?" November 2019, https://bdtechtalks.com/2019/11/20/social-artificial-intelligence/

[18] **M. Kaput,** "AI for social media: What you need to know," March 2020, https://www.marketingaiinstitute.com/blog/ai-for-social-media

[19] **M. Hogan,**" How artificial intelligence influences social media," May 2020, https://www.adzooma.com/blog/how-artificial-intelligence-influences-social-media/

[20] **Subbakrishna**, "5 Ways to harness the power of artificial intelligence in social media marketing," October 2017,
https://www.cloohawk.com/blog/5-ways-harness-power-artificial-intelligence-social-media-marketing?utm_content=buffer7447e&utm_medium=social&utm_source=twitter.com&utm_campaign=buffer

[21] **A. Yin**, "How we can use AI in digital marketing,"
https://www.motocms.com/blog/en/ai-in-digital-marketing/

[22] **J. Gupta**, "Role of artificial intelligence in social media," July 2020,
https://www.quytech.com/blog/role-of-artificial-intelligence-in-social-media/

[23] **M. Quadros**, "Artificial intelligence in social media marketing," September 2020,
https://www.socialbakers.com/blog/ai-in-social-media

[24] **M. Chui et al.**, "Applying artificial intelligence for social good," November 2018,
https://www.mckinsey.com/featured-insights/artificial-intelligence/applying-artificial-intelligence-for-social-good

[25] "AI in social media market by technology,"
https://www.marketsandmarkets.com/pdfdownloadNew.asp?id=92119289

[26] **C. Zilles**, "Taking advantage of AI in social media marketing," February 2019,
https://socialmediahq.com/taking-advantage-of-ai-in-social-media-marketing/

[27] **J. Hendler** and **A. M. Mulvehill**, Social Machines: The Coming Collision of Artificial Intelligence, Social Networking, and Humanity. Apress, 2016.

[28] **S. Natale**, Deceitful Media: Artificial Intelligence and Social Life After the Turing Test. New York: Oxford University Press, 2021.

CHAPTER 15
AI IN GAMING

"Failure doesn't mean the game is over, it means try again with experience."- *Anonymous*

15.1 INTRODUCTION

Artificial intelligence (AI) is the intelligence exhibited by an artificial entity, generally assumed to be a computer. It has been involved with gaming since day its inception. It is progressively being widely used in the gaming industry. AI in games is commonly used for creating player's opponents. It is the foundation of all video games. Games like Nim, checkers, or chess took advantage of smart algorithms to beat human players.

The games industry is one of the most lucrative industries due to the billion-dollar sales of digital games. The motivations for playing digital games are varied and different for different age groups. People play digital games for several reasons, from entertainment to professional training [1].

Game developers have been programming software to both pretend like it is a human. They become especially adept at using traditional techniques to achieve the illusion of intelligence. They have used AI to create art for games and push automated game design to new heights [2]. Their intent has not been to try and achieve some unprecedented level of human-like intelligence, but to create an experience that stimulates players in ways only the real world used to be capable of. The goal is to make the AI more human or at least appear to be so. Figure 15.1 shows a robot play a football game [3].

Figure 15.1   A robot plays a football game [3].

The origin of the application of AI in gaming can be found in the chess games between the computer IBM AI known as Deep Blue and the Russian master Gary Kasparov in 1996. In 2016, a Google AI system AlphaGo defeated top ranked player Lee Sedol in a game match of the Chinese board game Go. These examples suggest that AI systems can be dominant in just about any kind of game we humans can imagine. Gaming and AI have been conjoined intimately for nearly 70 years. For example. AI has been featured in genres like strategy games, shooting games, and even racing games.

This chapter provides an introduction on the applications of AI in different games. It begins with a brief review of AI. It explains the role that AI plays in gaming. It discusses some applications of AI in gaming. It provides some examples of game AI. It highlights the benefits and challenges of AI in gaming. It addresses how AI is being applied in gaming around the world. It presents the future of AI in gaming. The last section concludes with comments.

15.2 REVIEW ON ARTIFICIAL INTELLIGENCE

Artificial intelligence (AI) refers to computer systems that mimic human cognitive functions. It is a field of computer science that deals with intelligent machines. The term "artificial intelligence" was first used at a Dartmouth College conference in 1956. The main goal of AI is to enable machines to perform complex tasks that typically require human intelligence [4]. In simple terms, AI attempts to clone human behavior. An important feature of AI technology is that is can be added to existing technologies. AI is now one of the most important global issues of the 21st century. It is poised to disrupt our world and change processes and developments in the fields of science, engineering, education, business, entertainment, and agriculture.

The concept of AI is an umbrella term that encompasses many different technologies. AI is not a single technology but a collection of techniques that enables computer systems to perform tasks that would otherwise require human intelligence [5]. The major disciplines in AI include:

- Expert systems
- Fuzzy logic
- Neural networks
- Machine learning (ML)

- Deep learning
- Natural Language Processors (NLP)
- Robots

These AI tools are illustrated in Figure 15.2 [6].

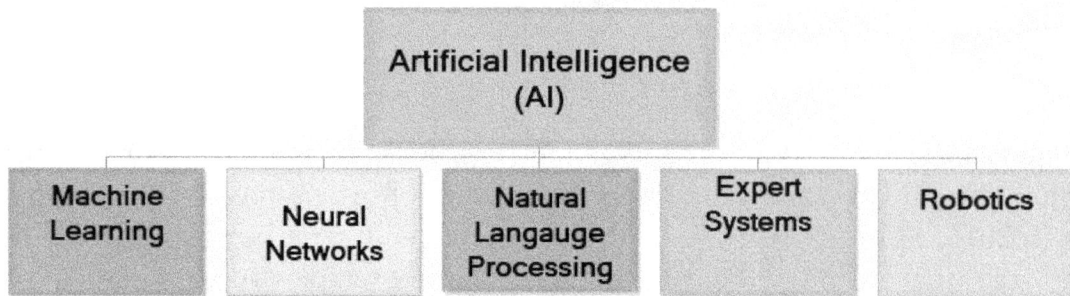

Figure 15.2 Different branches of Artificial intelligence.

Each AI tool has its own advantages. Using a combination of these models, rather than a single model, is recommended. AI systems are designed to make decisions using real-time data. They have the ability to learn and adapt as they make decisions. Game playing has been an area of research in AI from its inception. AI has been the backbone for countless aspects of computing and gaming. Machine learning can have a huge impact on game development because it addresses two problems: (1) Playing the game against (or alongside) human players, (2) Helping build the game dynamically for players [6,7].

15.3 AI IN GAMING

AI in a game can simply act as the player. The objective of having AI in games is to leave player feeling as though he is playing against real people. The player learns to think either strategically, tactically, or reactively. The player can give his squad two kinds of orders: explicit and implicit. Most games support only explicit orders: move, attack, guard, build, etc. Unlike explicit orders, implicit orders transmit information from the player to the units and assists them in making better autonomous decisions. To influence the player to perceive the creatures as intelligent, he has to be provided more insights on their actions, intentions, thoughts, and emotions (such as joy, fear, are trust, surprise, fear, disgust, and anticipation), which are simple to model. Autonomous behavior is hard to program manually, but it can be taught by providing examples [8].

Game playing has been an active research area in AI from the beginning. Artificial intelligence can be used in games in various ways. AI tools are used different assets of a game. AI can mimic, imitate, learn, forget, teach, and collaborate. It could be a testing tool to make your code or design more robust. It may be the unseen hand directing the whole affair. AI creates entirely new elements for the game — new levels, new rules, new environments. AI techniques can help generate intelligent, responsive behavior that molds on your reactions as a player. AI makes the game more interactive by boosting player's experience. They can adjust parameters such as speed and time.

While you are playing the game, the game is also playing you, exerting a bilateral effect. AI is more geared towards automation. In order to give the player non-human opponents, AI is needed in almost all games. AI-based games make you feel like you are

272

playing against another person. You will not need other human interactions when you play some of the multiplayer video games.

Game AI techniques often fall into two categories: deterministic and nondeterministic. Deterministic behavior or performance is predictable, fast, easy to implement, understand, test, and debug. Deterministic AI methods are the bread and butter of game AI. AI-based games are based on a finite set of actions or reactions whose sequence can be easily predicted by expert players. Nondeterministic behavior has a degree of uncertainty and is somewhat unpredictable. Nondeterministic techniques also can learn and extrapolate on their own [8].

Designers usually identify four main player modeling subtasks that are particularly relevant for game AI [9]:

- Progress a smart and human-like NPCs (Non-Person Character) to improve interaction with gamers.
- Predict human players' actions that lead to enhanced game testing and game design.
- Classify their behaviors to allow the personalization of the game.
- Discover frequent patterns or orders of actions to regulate how a player performs in a game.

15.4 APPLICATIONS OF AI IN GAMING

Modern games have advanced in multiple ways over the past decades. AI technologies such as machine learning, deep learning, neural networks, and natural language processing can produce high-quality video game and make modern games look amazing. Figure 15.3 shows a typical AI game [10].

Figure 15.3 A typical AI game [10].

The main goal of using AI in gaming is to give the player a realistic experience while playing even on a virtual platform. Some straightforward applications of AI in gaming include AI games, video games, virtual reality games, augmented reality games, strategic games, role playing games, online casinos, and mobile games.

AI Game: AI can be used in developing all types of games, especially for Game-AI. The Mind Game (renowned AI game) is viewed as a distinct subfield of AI and is often asked to solve fairly complex problems. Game AI is regarded as weak AI because it involves a broader range of purposes and technologies to give machines specialized intelligent qualities. In order to deliver an entertaining game for the user, the seven objectives of game AI objectives are [11]:

1- No clear cheating: Game AI has to be active without deceit.
2- No predicting behavior: Game AI behavior should not be predictable by the user.
3- No obvious inferior behavior: Game AI should not exhibit obvious inferior behavior against the user, so that it cannot be defeated easily by the user.
4- Use of the environment: Game AI has to exploit in a smart way the characteristics of the game environment.
5- Self-correction: Game AI should be capable of correcting its behavior in order to avoid repeating mistakes and aim at out-performing the rival.

6- Creativity: Game AI should be capable of generating novel solutions to unforeseen game circumstances.

7- Human-like behavior: The behavior exhibited by game AI should be equivalent in complexity to the human behavior.

Video Games: Artificial intelligence has been an integral part of video games since the 1950s. If you have played a video game, you have interacted with AI. Various video games, whether they are racing games, shooting games, or strategy games, have numerous features that are affected by AI. In video games, AI is used to generate responsive behaviors in non-player characters similar to human-like intelligence. Video game AI has revolutionized the way humans interact with all forms of technology. As far as video games are concerned, AI may be regarded as the set of techniques used to design the behavior of the "Non-Playable Characters" (NPC). In most video games, NPCs' behavior patterns are programmed and cannot learn anything from players. The main component of AI techniques that is widely used in video games is machine learning or more specifically, reinforcement learning. Game inventors dream of building video games and provide machine learning with a flexible environment for quick changes and easy customization. The most widely used AI technique in games is cheating. In AI-based video games, cheating refers to the programmer giving agents actions and access to information that would be unavailable to the player in the same situation [3]. Instead of learning how best to beat human players, AI in video games tends to enhance human players' gaming experience.

VR Games: Virtual reality (VR) is the simulation of a real environment using visual, auditory, and other stimuli. It involves using computer technology to create a simulated environment [12]. The common method of participating in VR is through a headset. Virtual reality game is a niche category when compared to the rest of the gaming industry. Machine learning is used in the video game industry, especially in virtual reality. VR is the future of gaming. VR games (or even just regular console games) will become more immersive and dynamic. Figure 15.4 illustrates an example of virtual reality game that brings people together [13].

Figure 15.4 An example of virtual reality game that brings people together [13].

Big tech companies like Facebook, Google, Microsoft, and Sony have greatly invested in developing VR hardware and games. These companies are busy making VR more consumer friendly.

AR Games: Augmented reality (AR) is a variation of VR. It plays a supplemental role rather than a replacement of reality. Typical AR devices include mobile phones and specially made of glasses. The AR technology powered Pokémon Go. It took a well-established brand (Pokémon) to get consumers to give it a try. AR is taking off faster than VR because people have an appetite for games that interact with reality, not remove them from it.

Online Casinos: AI is being used in several ways by online casinos. Online gambling platforms have started to use AI to collect smart data from players. AI is also used for improving the customer service experience for users. AI is used for slot machines to ensure that online casinos can minimize the chances that people will be able to cheat the device and create an unfair advantage for themselves. AI helps casino businesses to operate more efficiently [14].

Role Playing Games: In this kind of games, a player can play different types of human characters, such as a combatant, conjurer, or thief. The player does various kinds of activities in an extended virtual world. AI is implemented to take control over enemies similar to action games. These days major AI research areas on these types of games is to provide human interaction, social intelligence, and natural language interfaces to these support characters [15].

Strategy Games: In strategy games, the human controls various kinds of entities in order to conduct a battle from a god's eye view against one or more opponents. Strategy games cast a player in charge of a range of military units. The human is faced with problems of resource allocation, scheduling production, and organizing defenses and attacks. Strategy games on the market today are an even mix between mythical, fantasy, and science fiction campaigns, and recreations of historical battles. AI in strategy games needs to be applied both at the level of strategic opponents and at the level of individual units [15].

Mobile Games: Mobile gaming represents a killer application that is attracting millions of subscribers around the world. A crucial aspect of commercial success of a game is ensuring an appropriately challenging AI algorithm against which to play. Companies are already rolling out 5G for mobile devices, which make data available quickly, enable you to pull up an AR game, look through your screen, and get data on the world around you.

Other types of applications of AI in games include war games, football game, card games, shooting games, tactical games, simulation/God games, serious games, racing games, online games, adaptive games, commercial games., and nonplayer character (NPC).

15.5 EXAMPLES OF GAME AI

The following games employ AI and are interesting, clever, and novel [16]:

(1) The Division's AI-driven path finding for changing cover. AI can drive mechanics that help the player get around faster.

(2) Forza's Drivatar adaptive AI system. AI that mimics real people can make enemies and opponents seem more human.

(3) Alien: Isolation's hunting Xenomorph. An enemy AI designed to relentlessly hunt the player as they roam about the game world.

(4) The Ice-Bound Concordance's combinatorial narrative. You can use AI to tell a dramatically satisfying story.

(5) City Conquest's playtesting via genetic algorithm. AI can help you make your game better by playtesting to find dominant species.

(6) Testing for walkability in The Witness. AI can do the grunt work for you in finding all the nasty problems that could frustrate players simply trying to explore your game's world.

(7) The AI Directors in Left 4 Dead, Rocksmith 2014, and others. Every player is different, and AI can help ensure that everybody gets a satisfying, challenging experience.

(8) With FEAR, the AI works in a way that it could easily fool the player into thinking they were facing a group of human players. FEAR's enemies are skilled commandos trying to put a stop to your mission.

(9) AI project called Angelina, developed by Michael Cook in 2011, is capable of designing video games from scratch in a way that is surprisingly simple.

(10) Grand Theft Auto V is the most profitable and perhaps the most successful entertainment product of all time at $6 billion in revenue, eclipsing all movies, TV shows, and music in terms of total revenue.

15.6 BENEFITS

Gaming is the future of entertainment. In essence, games are learning devices. AI is initiating a new era of smart video games. AI essentially consists of algorithms which you can tame whichever you want. AI has a rich history and has been the backbone for countless aspects of computing and gaming. AI serves to improve the game-player experience. AI can play multiple roles in gaming. AI can also be used to enhance existing games. "Video games offer the best test of intelligence we have. Combining AI with virtual or augmented reality opens the gates to add reality factor to video games. Other benefits include the following.

Better Games: AI-based games are better and more exciting than traditional games. The days of traditional gaming are gone. Today, game lovers expect more from their games. AI improves the gaming experience by making games smarter, more realistic, and just all-around better.

Human Enjoyment: Human derive entertainment from games. Enjoyment of games is derived from enjoying progress, mastery, proficiency, experimentation, and learning. Players will enjoy playing against/ along with the computer because of each character's intelligence. Human enjoyment of games is derived from enjoying progress, mastery, proficiency, experimentation, and learning. This aspect of human nature goes all the way back to Roman coliseum, where actual gladiators and wild animals were employed.

Indefatigable Opponent: AI in games takes the role of a never-bored and never-boring opponent. The games are designed to rescue the player from the boredom of repetition and letting him focus just on the interesting aspects of the game.

A Boom: While AI in some form has long appeared in video games, it is considered a booming new frontier in how games are both developed and played.

Cheating AI: Perhaps the most widely used AI technique in games is cheating. Although AI already can beat human players, it can cheat. "Cheating" simply implies that an AI-controlled character makes the moves that are not possible with a human player.

Retaining Users: A critical challenge for gaming companies is to attract and retain users. AI offers unique retention by providing immersive experiences apart from unique perspectives to engage gamers.

15.7 CHALLENGES

AI is not yet capable of creating entire high-quality games from scratch. Games can be addictive to the player. Elon Musk has recently warned the world that the fast development of AI with learning capability by Google and Facebook would put humanity in danger. The gaming industry is conservative, and publishers or game makers need to take risks. There is the temptation of preferring to keep doing that same thing. It is difficult to create thoroughly robust AI because its development is constrained to the scope of an individual game project. Other challenges include the following.

Complexity: The major limitation to strong AI is the depth of thinking and the extreme complexity of the decision-making process. Game AI algorithms are used in a disparate field inside a game.

Cost: Perhaps the only barrier to fully utilizing AI technology in gaming is the eventual limit of money and time. A related challenge is the cost incurred in the maintenance and repair.

Fear: The idea of machines replacing human beings sounds threating. If robots begin to replace humans in every field, it will eventually lead to unemployment.

Emotion: In current computer games, emotion is either non-existent or superficial.

15.8 GLOBAL AI IN GAMING

Computer games constitute a major branch of the entertainment industry worldwide. In recent years, AI has become one of the hottest buzzwords in the gaming industry. AI will keep on revolutionizing the video gaming and e-gaming business around the world. This section examines the role of AI in gaming in different nations.

United States: According to the Entertainment Software Association, there are 155 million U.S. Americans that play videogames, with an average of two gamers in each game-playing U.S. households. This shows the good condition of the video game industry, which has taken the lead in entertainment industry. This situation has been a motivation for the research applied to video games. The computer game industry has grown rapidly over the last decade. Currently computer games enterprise is the biggest sector of the entertainment industry in the US.

United Kingdom: UK based start-up Sonantic has developed an artificial voice technology, a kind of virtual actor, which can deliver lines of dialogue with convincing emotional depth, adding fear, joy and shock, depending on the situation. The system requires a real voice actor to deliver a couple of hours of voice recordings, but then the AI learns the voice and can perform the role itself. Figure 15.5 illustrates the UK game industry size comparison.

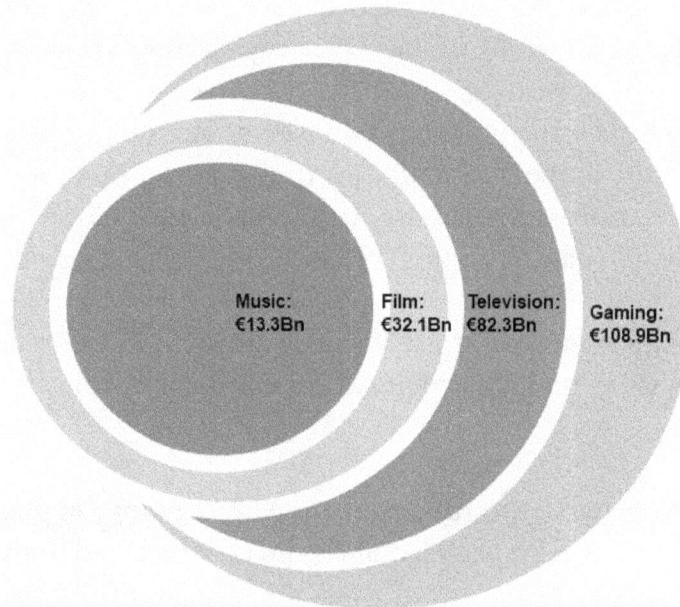

Figure 15.5   UK game industry size comparison.

India: India has the cheapest mobile data globally and that makes it even more favorable for the eSports industry. Analytics India Magazine got in touch with Tarun Gupta, the Founder of Ultimate Battle, to understand how technology determines the growth of the eSports industry and how it is leveraging deep-tech to navigate through the times of pandemic. The massive usage of AI helps in delivering top quality eSports content to a massive gaming audience worldwide.

15.9 THE FUTURE OF AI IN GAMING

The main objective of game developers in the future is to plan strong frameworks inside games. This will require present-day AI techniques and AI-based voice intelligence, which is now changing the way in which games are played.

The future of the application of AI technology lies in the development of video games and the ability of the technology to increase the human connection, i.e., AI that is human-like, emotional, and responsive. AI is clearly the future of gaming and the future of AI in video games would naturally point to automation. In the future, AI becomes a kind of collaborator with humans, helping designers and developers create art assets, design levels, and even build entire games from the ground up. Big tech companies such as Sony, Nintendo, Microsoft, Apple, Google, and Amazon are seizing the moment and developing gaming products.

15.10 CONCLUSION

Artificial intelligence is rapidly changing the gaming industry. Games have been regarded as the perfect testbed for AI techniques. Modern computer games often feature realistic environments by employing 3D animated graphics to give the impression of reality. State-of-the-art games can recreate real-life environments with a surprising level of detail. The demands of the gaming community and the games themselves keep evolving.

AI is evidently the future of gaming. It has started to truly take center stage in the next generation of games. It will continue to have a great impact on the game development as well as game industry. For more information about artificial intelligence in games, one should consult the books in [8, 17-26] and related journals:

- AI Time Journal
- International Journal of Computer Games Technology
- IEEE Transactions on Computational Intelligence and AI in Games

REFERENCES

[1] **M. N.O. Sadiku, S.M. Musa,** and **R. Nelatury**, "Digital games," International Journal of Research and Allied Sciences, vol. 1, no. 10, Dec. 2016, pp. 1,2.

[2] **N. Statt**, "How artificial intelligence will revolutionize the way video games are developed and played," March 2019, https://www.theverge.com/2019/3/6/18222203/video-game-ai-future-procedural-generation-deep-*learning*

[3] "Artificial intelligence in video games," *Wikipedia,* the free encyclopedia https://en.wikipedia.org/wiki/Artificial_intelligence_in_video_games#:~:text=In%20video%20games%2C%20artificial%20intelligence,similar%20to%20human%2Dlike%20intelligence.&text=It%20serves%20to%20improve%20the,machine%20learning%20or%20decision%20making.

[4] **S. Greengard**, "What is artificial intelligence?" May 2019, https://www.datamation.com/artificial-intelligence/what-is-artificial-intelligence.html

[5] https://in.pinterest.com/pin/828662400161409072/

[6] M. **N. O. Sadiku, S. M. Musa**, and **A. Ajayi-Majebi,** "Artificial intelligence in gaming," International Journal of Trend in Scientific Research and Development, vol. 5, no. 2, Jan.-Feb. 2021, pp. 710-714.

[7] **J. Stephenson**, "6 Ways machine learning will be used in game development," November 2018, https://www.logikk.com/articles/machine-learning-in-game-development/ https://www.logikk.com/articles/machine-learning-in-game-development/

[8] **G. Seemann** and **D. M. Bourg**, AI for Game Developers. O'Reilly Media, 2004, chapter 1.

[9] **S. Ghosh**, "AI games – Video games with artificial intelligence," https://www.minditsystems.com/ai-games-video-games-with-artificial-intelligence/

[10] "Artificial intelligence in games," September 2018, https://medium.com/aifrontiers/an-overview-of-artificial-intelligence-for-video-games-f491229c0e7d

[11] **H. K. Barznji**, "Artificial intelligence and game development," https://www.researchgate.net/publication/330290637_Artificial_Intelligence_and_Game_Development

[12] M. **N. O. Sadiku**, **K. G. Eze**, and **S. M. Musa**, "Virtual reality: A primer," International Journal of Trend in Research and Development, vol. 7, no. 2, March-April 2020, pp. 160-162.

[13] **H. Koss**, "What does the future of the gaming industry look like?" May 2020, https://builtin.com/media-gaming/future-of-gaming

[14] **"The rise of AI in gaming,"** December 2020, https://www.analyticsinsight.net/the-rise-of-ai-in-gaming/#:~:text=AI%20improves%20the%20overall%20gaming,to%20understand%20the%20action%20better.

[15] **"Applications of artificial intelligence to game design,"** https://howtowrite.customwritings.com/post/applications-artificial-intelligence-game-design/

[16] **R. Moss**, "7 Examples of game AI that every developer should study," April 2016, https://www.gamasutra.com/view/news/269634/7_examples_of_game_AI_that_every_developer_should_study.php

[17] **I. Millington** and **J. Funge**, Artificial Intelligence for Games. Boca Raton, FL: CRC Press, 2nd edition, 2009.

[18] **G. N. Yannakakis** and **J. Togelius**, Artificial Intelligence and Games. Springer, 2018.

[19] **J. Togelius**, Playing Smart: On Games, Intelligence, and Artificial Intelligence. Cambridge, MA: The MIT Press, 2019.

[20] **D. Charles et al.**, Biologically Inspired Artificial Intelligence for Computer Games. London, UK: IGI Global, 2007.

[21] S. **Rabin (ed.),** AI Game Programming Wisdom. Charles River Media; 2002.

[22] **S. Cossu**, Beginning Game AI with Unity: Programming Artificial Intelligence with C#. Apress, 2021.

[23] **J Togelius**, Playing Smart: On Games, Intelligence, and Artificial Intelligence. MIT Press, 2019.

[24] **P. A. González-Calero** and **M. A. Gómez-Martín**, Artificial Intelligence for Computer Games. Springer, 2011.

[25] J. **Schaeffer,** and **J. van den Herik (eds.)**, Chips Challenging Champions: Games, Computers and Artificial Intelligence. Elsevier Science, 2002.

CHAPTER 16
HUMANIZED AI

"The thing that's going to make artificial intelligence so powerful is its ability to learn, and the way AI learns is to look at human culture." - *Dan Brown*

16.1 INTRODUCTION

Intelligence refers to the mental ability for reasoning, problem-solving, and learning. It has the capabilities to learn from experience, adapt to new situations, and handle abstract ideas. It can be measured by standardized tests and used to determine educational achievement, job performance, and health condition. Artificial Intelligence (AI) is intelligence demonstrated by a machine, as opposed to human intelligence. The field of AI is based on the premise that human intelligence can be so precisely described and simulated by a machine. Evidence of the positive impact of AI systems is all around us. Organizations and individuals around the world are creating core principles around AI with an emphasis on a more humanist approach. They realize the significant advantages AI can bring to their business [1].

Human beings have the natural ability to sense their environment, interpreting what they see, responding to stimuli, and empathizing with the environment and with other fellow human beings. However, humans are limited in terms of computing high number and in repetitive tasks. AI can extend the skills of humans by learning to perform and automate tasks in manners designed by humans. AI performs tasks in a robotic manner. It will never be possible for such machines to completely replace the human beings partly due to the fact that AI lacks a human touch. Although AI can provide massive computational powers, it is yet to handle the concept of emotions like happiness, sadness, depression, stress, anger, and pain. This is evident in the case of wealth management where a human expert, rather than an AI counterpart, is received well by the customer. However, AI improves its ability to empathize only with the availability of data. It requires a lot of intimate information about an individual to be able to truly understand the person's physical or mental limitations. Research on AI employs different tools from many fields, including computer science, psychology, philosophy, logic, neuroscience, cognitive science, linguistics, operations research, economics, control theory, and probability [2,3].

This chapter provides an introduction to humanized AI. It begins by briefly reviewing AI. It describes the concept of humanized AI. It presents some applications of humanized AI. It highlights the benefits and challenges of humanized AI. It explains how humanized AI is implemented around the world. It presents the future of humanized AI. The last section concludes with comments.

16.2 REVIEW ON ARTIFICIAL INTELLIGENCE

AI refers to computer systems that mimic human cognitive functions. It is a field of computer science that deals with intelligent machines. AI has long history which is actively and constantly changing and growing. The term "artificial intelligence" was first used at a Dartmouth College conference in 1956. The main goal of AI is to enable machines to perform complex tasks that typically require human intelligence [4]. In simple terms, AI attempts to clone human behavior. An important feature of AI technology is that is can be added to existing technologies. AI is now one of the most important global issues of the 21st century. It is poised to disrupt our world and change processes and developments in the fields of science, engineering, education, business, entertainment, and agriculture.

The concept of AI is an umbrella term that encompasses many different technologies. AI is not a single technology but a collection of techniques that enables computer systems to perform tasks that would otherwise require human intelligence [5]. The major disciplines in AI include:
- Expert systems
- Fuzzy logic
- Neural networks
- Machine learning (ML)
- Deep learning
- Natural Language Processors (NLP)
- Robots

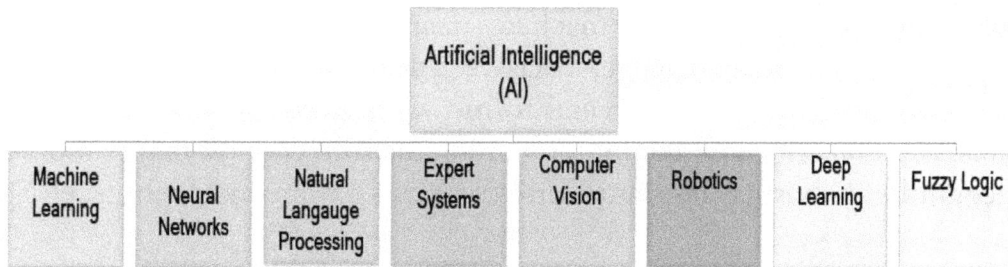

Figure 16. 1 Different branches of artificial intelligence.

Each AI tool has its own advantages. Using a combination of these models, rather than a single model, is recommended. AI systems are designed to make decisions using real-time data. They have the ability to learn and adapt as they make decisions. Figure 16.2 shows the evolution of artificial intelligence.

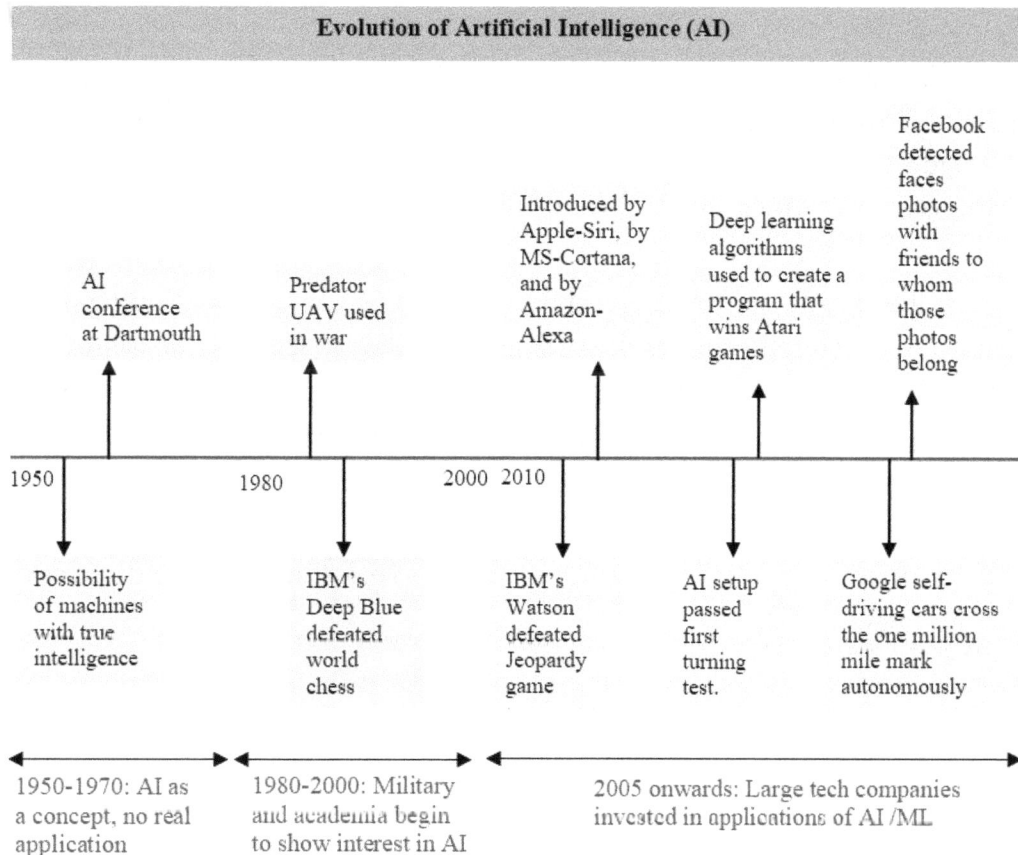

Figure 16.2   Timeline evolution of artificial intelligence.

AI has benefited many areas such chemistry and medicine, where routine diagnoses can be initiated by AI-aided computers. It embraces a wide range of disciplines such as

287

computer science, engineering, machine learning, chemistry, biology, physics, astronomy, medicine, neuroscience, social sciences, and the military. Today, AI is integrated into our daily lives in several forms, such as personal assistants, automated mass transportation, aviation, computer gaming, facial recognition at passport control, voice recognition on virtual assistants, driverless cars, companion robots, etc. [6].

16.3 CONCEPT OF HUMANIZED AI

Humanized AI attempts to create AI that is more human. It is a call for building human systems where the efficiency, programs, processes, and outcomes can be assessed and improved on the basis of data. It envisions the state of the world where humans and machines work together, each leveraging its comparative advantages with humans in the drivers' seat. Various attempts have been made to integrate fundamentally human ideas like judgment, empathy, or fairness into an AI equation.

Whatever affects humans requires a humanized approach. AI is no exception. Humanized AI is that which understands human emotions like happiness, stress, urgency, anger, to detect emotions like laughter, anger, arousal, and pain. It responds to natural language very much like a human friend.

Perhaps a good way to understand humanized AI is to first understand human intelligence. Human intelligence is the mental capacities to learn, understand, and reason, including the capacities to comprehend ideas, plan, solve problems, and use language to communicate. It is the mental quality that consists of the abilities to learn from experience, adapt to new situations, understand abstract concepts, and use knowledge to adapt to the environment. It integrates cognitive functions such as perception, attention, memory, language, and planning. Figure 16.3 illustrates the three components of human intelligence [7].

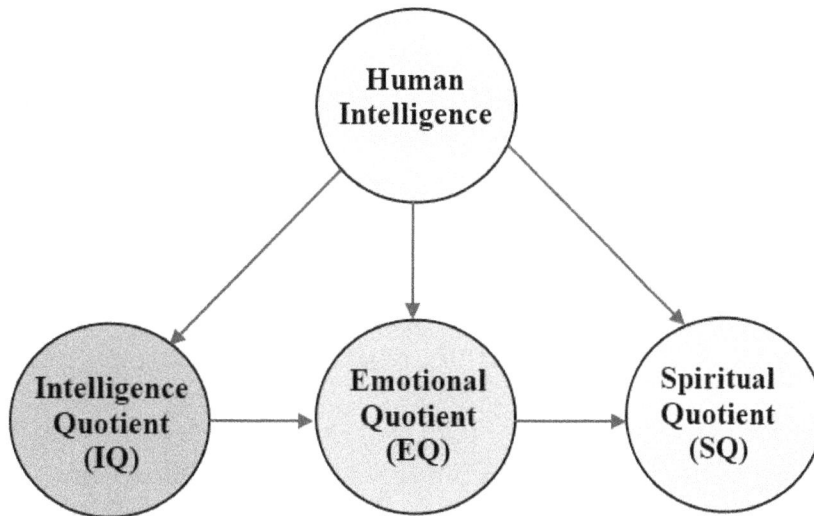

Figure 16.3 Three component of human intelligence.

Humans have the following characteristics [8]:

1. **Limited memory:** Humans have limited memory, and sometimes, even when faced with the exact same situation than previously encountered. They use memory and thinking, while robots use the built-in instructions, designed by engineers and scientists.

2. **Limited processing:** When problem size grows, strong rational skills are required, and humans are limited. However, human beings can work on multiple responsibilities.

3. **Emotions:** Humans have emotions. Emotions are probably the most impacting factor for human decision-making.

4. **Environment: To feel human involves feeling secure and familiar with our surroundings, both physical and cultural.** Humans have the ability to change their own environment using the knowledge gained. They can adapt to the environment using a combination of several cognitive processes. Social context often defines appearance through cultural influences.

5. **Creativity:** Creativity and imagination are main characteristics of human species. Human creativity is superior to AI because it is the creation of an intelligent Designer, God. AI is artificial, little, and temporary created by humans.

In contrast to human intelligence, artificial Intelligence has the following advantages [9]

- **Speed of execution:** For example, a doctor can make a diagnosis in 10 minutes, AI system can make a million for the same time.

- **Less Biased: AI systems** do not involve emotions or biased opinions on decision making process.

- **Operational Ability:** They do not expect halt in their work due to saturation.

- **Accuracy:** Preciseness of the output increases.

Artificial intelligence can beat human intelligence in some specific areas. Let us consider three examples. First, in chess a supercomputer has beaten the human player. Second, AI technologies such as IBM Watson are being used at some medical centers to support diagnosis and help man. Third, Sophia is social robot created by former Disney Imagineer David Hanson. Sophia behaves like human. The robot has sat for TV interviews, appeared on the cover of ELLE magazine, been parodied on HBO, and was appointed the UN's first non-human "innovation champion." The guiding principle of AI is not to become autonomous or replace human intelligence but bolster the latter.

16.4 APPLICATIONS

Humanized AI is that which understands human emotions like joy, stress, urgency, anger and pain when humans display them through speech, facial and physical expressions. Humanization reflects on those things which make us feel more human. Humanizing principles can be applied to every machine that involves human-AI collaboration [10]. Common areas in which humanized AI is currently observed include education, healthcare, and financial services which affect human life directly, socially, physically, and economically [11].

Healthcare: One of the greatest challenges faced by healthcare professionals is providing humanized care that is acceptable in the current medical practices. Using neural network, AI algorithm can identify and categorize the level of depression through a study of characteristics of speech, amount of breath, word choice, etc. Healthcare professionals should rethink their behaviors when providing care to patients so they can ensure quality care based on humanization [12]. Figure 16.4 shows an example of Human-AI system in healthcare.

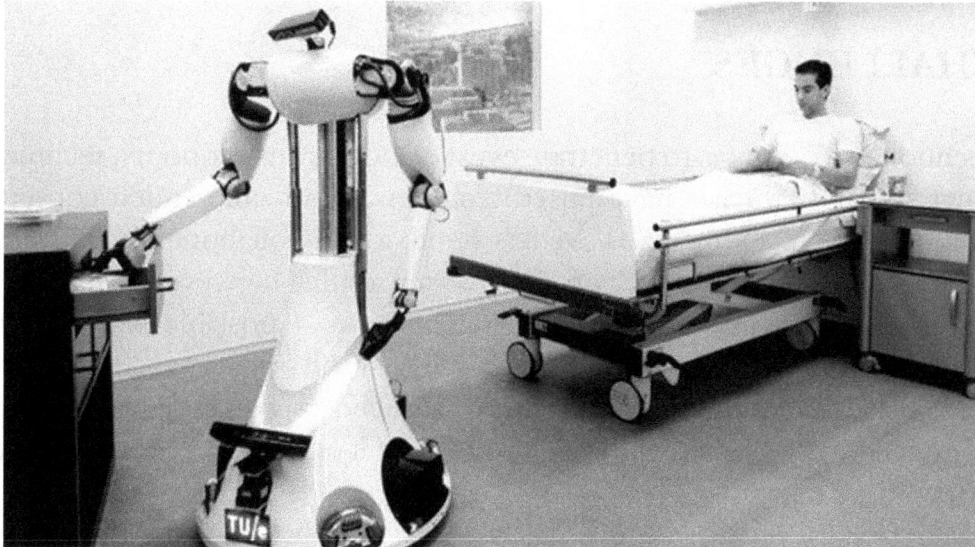

Figure 16.4   An example of human-AI system in healthcare [13].

Financial Services: Several financial multi-national companies are experimenting with AI in finance to bring more benefits to customers. Financial services industries have tended to build more secure systems as people's money is at stake. Companies are using AI principles to leap ahead on innovation and profitability.

Education: AI and personalization go hand in hand. Personalized learning through analytics and AI has been enabled for several classrooms by pioneers. AI can be used to engage with the students in solving problems. Education will be crucial in the 21st century and may need to be redesigned to meet the challenges.

16.5 BENEFITS

A lot of companies benefit from how powerful and scalable their AI is. AI is capable of fighting fraud which affects consumers. AI systems have significantly taken over many tasks, especially when it comes to monotonous judgments. The utilization of AI will make life more convenient for humankind. AI systems would be able to express a mixture of feelings, such as fear, doubt, hope, anger, revenge, trust, mistrust, empathy, etc., as an integral part of human behavior.

16.6 CHALLENGES

As AI technology continues to penetrate every sector of our economy, technical experts, policy analysts, and ethicists have raised concerns about unintended consequences of widespread adoption. Humanized AI has limitations. Collaboration between humans and machines does not necessarily produce better results. However, some people regard the collaboration between humans and machines as a possible way to mitigate the problems of bias in machine learning. For the sake of privacy, security, transparency, unfairness, bias, discrimination, and ethics, we need to be thoughtful about how AI technology can be used, where it can be used, when it can be used, and what it can access [14].

As AI technology continues to develop, researchers must ensure that AI-enabled systems are governable, open, transparent, and understandable; Some are concerned about the fate of humankind in the "fourth industrial revolution," and the emergence of big data and AI. Human-AI collaboration has cultural and social implications. For example, should the driverless car kill an elderly people if it has no choice? Today's AI systems are still dependent on human input and not completely autonomous. It is crucial for an AI to simulate human-like learning, decision making and, most importantly, autonomy [15]. This will increase confidence in their decision-making. Other challenges include [16]:

- **Privacy and Protection:** Privacy has been a major concern for government, business, academia, and civil society organizations. Privacy and data protection is very vital to human rights principles. Legal scholars believe that AI poses huge privacy and data protection challenges. These include informed consent, surveillance, infringement of data protection rights of individuals.

- **Ethical Issues:** The more we train machines and expect them to emulate us as social beings, the more we are confronted with the same ethical issues that accompany human interactions.

- **Legal issues:** There are a number of legal issues and human rights challenges related to AI. AI entities should be regarded as liable, i.e., responsibility for harm caused by the manufacturer, end-user or owner.

- **Vulnerability:** This refers to the state of being exposed to the possibility of being attacked or harmed. People who are prone to being harmed, exploited or discriminated include children, women, the elderly, people with disabilities, and members of ethnic minority groups. Cyber security vulnerabilities also pose a significant threat.

- **Job Loss:** The growing use of AI technology will cause a widespread automation, reducing the need for human workers. As AI enhances human effectiveness, it also threatens human autonomy and capabilities. However, machines are still far from what human brains are capable of doing. Figure 16.5 depicts how AI will make the workplace more human, not less [17].

Figure 16.5   Artificial intelligence will make the workplace more human, not less [17].

- **Lack of Trust:** We find it difficult to trust machines we do not really understand. This lack of trust makes users to treat AI like modern witchcraft or as a black box. To increase trust in AI systems, researchers are exploring the concept of "explainable AI" by showing humans what is really going on.

Not addressing these issues will hinder the adoption of AI.

16.7 GLOBAL HUMANIZED AI

Artificial intelligence continues to find its way into our daily lives by using the GPS navigation and check-scanning machines. But experts have only started to look at the impact of AI on human rights worldwide. AI has the potential to increase inequality and generate economic disruption, social unrest, and even political instability in the developing world. UN human rights investigators should continue researching and publicizing the human rights impacts resulting from AI systems. Governments can play a more active role in advocating for AI development that respects human rights. There is enormous potential for how AI can benefit the developing world and what it can contribute towards achieving the UN's Sustainable Development Goals. The expected benefits promise to be transformative, but the negative repercussions could be magnified in developing countries, where the livelihoods of many people are precarious and social institutions can be fragile [18].

AI poses policy questions across a range of areas in international relations and security. Each national government has several roles to play. It can convene conversations about important issues and help to set the agenda for public debate. This will help create a wide range of approaches as they make decisions on AI policy. Here we consider how humanized AI is implemented in different nations [19].

United States: The US Department of Defense adopted a series of ethical principles for AI in an effort to prioritize national security. AI has a whole new dimension of an endless pool of opportunities for humans to explore. Stanford University researchers trained a deep neural network to "predict" the sexual orientation of their subjects, without obtaining consent, using a set of images collected from online dating websites. In 2017, New York City passed a law that aims to help ensure that algorithms used by city agencies are transparent, fair, and valid.

European Union: EU data protection law gives individuals rights to challenge and request a review of automated decision-making that significantly affects their rights. Data subjects have the right to object on grounds relating to their particular situation. The European Commission Expert Group on Liability and New Technologies concluded in its review of existing liability regimes on emerging digital technologies. The EU has demonstrated an interest in regulating technology companies with an appeal to rights-based principles. It has also produced its Declaration of Cooperation on Artificial Intelligence, Artificial Intelligence for Europe, Ethics Guidelines for Trustworthy AI, Policy and Investment Recommendations for Trustworthy Artificial Intelligence, and White Paper on Artificial Intelligence.

China: There are reports that the government of China is deploying systems to categorize people by social characteristics. This Social Credit System is being developed to collect data on Chinese citizens and score them according to their social trustworthiness, as determined by the government.

South Africa: In South Africa, a classification system built on databases that sorted citizens by racial taxonomies was deployed to implement the racist policies of the apartheid regime. The South African constitution, adopted in 1996, directly accounted for the discriminatory policies of the past. The constitution establishes equality, human dignity, and human rights as its legal foundations and core values.

16.8 FUTURE OF HUMANS AND AI

Experts say the rise of AI will make most people better off over the next decade. Some are concerned about how advances in AI will affect what it means to be human, to be productive and to exercise free will. Human rights alone cannot address all AI present and future challenges.

As emerging AI technology continues to spread, will people be better off than they are today? It is humans, not machines, that will build the future. Since AI is still in its developmental stage, the future lies in how well we humans incorporate human values and safety measures in AI systems [20]. We have attempted to build intelligence in machines to ease our work. There are bots, humanoids, robots, and digital humans that either outplay humans or collaborate with humans. These AI-driven applications have higher speed of execution, have higher operational ability and accuracy, while also highly significant in tedious and monotonous jobs compared to humans.

16.9 CONCLUSION

The advances in AI are reaching new peaks with unprecedented speed. As AI technologies continue to impact our daily lives, it is expected that AI systems will work synergistically with humans. Organizations and people around the world are building trustworthy and dependable AI. Business leaders are getting significant value from advanced AI in their companies. AI is critical to the business of the future.

Although experts claim that the rise of networked artificial intelligence will make most people better off with time, many are concerned about how advances in AI will affect what it means to be human, to be productive, and to exercise free will. The AI algorithms of machines simply extend human capabilities and not replace them. A human AI is a human social system that would apply and leverage the power of data and the principles of AI. It is hoped that human systems would become better off, safer, fairer, more civil, and more sustainable with time [14]. Intelligent machines are the future of humanity. For more information about human AI systems, one should consult the books in [18,21-24].

REFERENCES

[1] **M. N. O. Sadiku, Y. P. Akhare, A. Ajayi-Majebi,** and S.M. **Musa,**" Humanized artificial intelligence," International Journal of Advances in Scientific Research and Engineering, vol. 6, no. 12, December 2020, pp. 9-14.

[2] "What is artificial intelligence?" May 2020,

https://zaincomptech.blogspot.com/2020/05/what-is-ai.html

[3] **J. Shabbir** and **T. Answer,** "Artificial intelligence and its role in near future," Journal of Latex Class Files, vol. 14, no. 8, August 2015.

[4] **S. Greengard,** "What is artificial intelligence?" May 2019,

https://www.datamation.com/artificial-intelligence/what-is-artificial-intelligence.html

[5] https://in.pinterest.com/pin/828662400161409072/

[6] **Y. Mintz** and **R. Brodie,** "Introduction to artificial intelligence in medicine," Minimally Invasive Therapy & Allied Technologies, vol. 28, no. 2, 2019, pp. 73-81.

[7]https://www.researchgate.net/figure/Three-levels-of-Human-Intelligence-System_fig1_305265106

[8] **J. Lynden,** "What does it mean to 'humanize' technology?" November 2018,

https://becominghuman.ai/what-does-it-mean-to-humanise-tech-1c6c4f28bf91

[9] "No longer science fiction, AI and robotics are transforming healthcare,"
https://www.pwc.com/gx/en/industries/healthcare/publications/ai-robotics-new-health/transforming-healthcare.html

[10] "Differences between artificial intelligence vs human intelligence,"
https://www.educba.com/artificial-intelligence-vs-human-intelligence/

[11] **T. Bandopadhyay** and **S. Bharani**, "Humanized AI for analytics that matter," January 2019,
https://www.wipro.com/en-US/analytics/humanized-ai-for-analytics-that-matter/

[12] **E. Norton**, "The application of humanization theory to health-promoting practice," Perspectives in Public Health, vol. 135, no. 3, May 2015, pp. 133-137.

[13] **R. Garza-Hernandez et al.,** "Surgical patients' perception about behaviors of humanized nursing care," Hispanic Health Care International, July 2019.

[14] "Artificial intelligence examples,"
http://major.magdalene-project.org/artificial-intelligence-examples/

[15] **J. Kite-Powell**, "How do we create artificial intelligence that is more human?"
https://www.forbes.com/sites/jenniferhicks/2019/03/19/how-do-we-create-artificial-intelligence-that-is-more-human/#6787a1d14920

[16] **R. Rodrigues**, "Legal and human rights issues of AI: Gaps, challenges and vulnerabilities," Journal of Responsible Technology, vol. 4, December 2020.

[17] **P. Uria-Recio**, "Artificial intelligence will make the workplace more human, not less," August 2019,
https://towardsdatascience.com/artificial-intelligence-will-make-the-workplace-more-human-not-less-49af1ce6cd0d

[18] **S. Russell**, Human Compatible: Artificial Intelligence and the Problem of Control. Viking, 2019.

[19] **M. Latonero**, "Governing artificial intelligence: Upholding human rights & dignity," https://datasociety.net/wp-content/uploads/2018/10/DataSociety_Governing_Artificial_Intelligence_Upholding_Human_Rights.pdf

[20] **D. Bhushan**," Artificial Intelligence vs Human Intelligence: Humans, not machines, will build the future," February 2020, https://in.springboard.com/blog/artificial-intelligence-vs-human-intelligence/

[21] **A. Daniele** and **Y. Z. Son**, "AI+Art=Human," Proceedings of the 2019 AAAI/ACM Conference on AI, Ethics, and Society, Honolulu, HI, USA, January2019, pp. 155-161.

[22] **E. Topol**, Deep Medicine: How Artificial Intelligence Can Make Healthcare Human Again. Basic Books, 2019.

[23] **P. R. Daugherty** and **H. J. Wilson**, Human + Machine: Reimagining Work in the Age of AI. Harvard Business Review Press, 2018.

[24] **N. Schwarzer,** Humanized Artificial Intelligence: Spiritual Computers - Make Them Believe and They'll Start to Think. Unknown Publisher, 2017.

CHAPTER 17
WEARABLE AI

"I have no doubt that in the future, wearable devices like Fitbit will know my blood pressure, hydration levels and blood sugar levels as well. All this data has the potential to transform modern medicine and create a whole new era of personalized care." - *Michael Dell*

17.1 INTRODUCTION

Artificial intelligence (AI) is a branch of computer science that studies how machine intelligence imitates human behaviors and cognitive abilities. AI technologies have been successfully applied for wearable devices (WDs) and inspired various applications. A wearable device essentially consists of two different components: wearable and body sensors. It incorporates sensors, memory, solar cells, and batteries. It stays in contact with the body for extended periods of time. Traditional materials for wearables are mostly metals and semiconductors with relatively poor mechanical flexibility. Modern wearable technologies are characterized by body-worn devices, as smart clothing, e-textiles, and accessories [1]. Wearable devices can be used to collect various data to support a series of innovative applications. Wearable-AI has been widely applied in consumer electronics, food security, healthcare, enterprise, disease diagnosis, environmental monitoring, sports, and games.

Wearables and hearables are everywhere. They have become an inherent part of our daily lives. They are typical examples of intelligence devices that have been adopted recently and rapidly. Today, WDs have numerous applications due to their integration with AI. The integration of AI and wearable sensors enables better acquisition of patient data and improved design of wearable sensors. Different types of wearable AI devices include earwear, wristwear, eyewear, and other bodywear. They also include smart phones, smart watches and wristbands, smart glasses, smart clothes and socks, smart sneakers, and smart headphones. Wearable devices can be attached to shoes, eyeglasses, earrings, clothing, gloves, and wrist watches [2].

The growth in wearable AI market can be attributed to rising proliferation of advanced technologies such as AI and 5G smartphone, the rapid urbanization, and rising disposable income in developing nations. The increasing smartphone penetration is driving the adoption of consumer electronics products such as fitness and health monitoring smart wearables such as smartwatches and fitness bands.

The smart wearable device manufacturers such as Microsoft, Apple, Google, Fossil, Facebook, Fitbit, and Samsung are leveraging on various next-generation technologies in their product offerings. They are focusing on developing new smart wearables to gain a niche in the market [3].

This chapter focuses on the various applications of wearable-AI. It begins by briefly reviewing artificial intelligence. It presents different kinds of wearables and AI in wearables. It covers various applications of AI wearables. It highlights the benefits and challenges of AI wearables. It addresses global implementation of AI wearables and the future of AI in wearables. The last section concludes with comments.

17.2 REVIEW ON ARTIFICIAL INTELLIGENCE

Artificial intelligence (AI) refers to computer systems that mimic human cognitive functions. It is a field of computer science that deals with intelligent machines. AI has long history which is actively and constantly changing and growing. The term "artificial intelligence" was first used at a Dartmouth College conference in 1956. The main goal of AI is to enable machines to perform complex tasks that typically require human intelligence [4]. In simple terms, AI attempts to clone human behavior. An important feature of AI technology is that is can be added to existing technologies. AI is now one of the most important global issues of the 21st century. It is poised to disrupt our world and change processes and developments in the fields of science, engineering, education, business, entertainment, and agriculture. The concept of AI is an umbrella term that encompasses many different technologies. AI is not a single technology but a collection of techniques that enables computer systems to perform tasks that would otherwise require human intelligence [5]. The major disciplines in AI include:

- Expert systems
- Fuzzy logic
- Neural networks
- Machine learning (ML)
- Deep learning
- Natural Language Processors (NLP)
- Robots

These AI tools are illustrated in Figure 17.1.

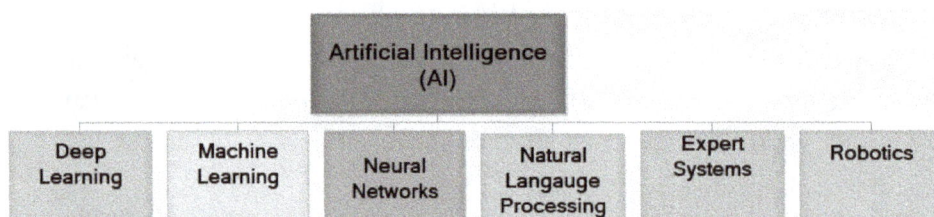

Figure 17.1 Different branches of artificial intelligence.

Each AI tool has its own advantages. Using a combination of these models, rather than a single model, is recommended. AI systems are designed to make decisions using real-time data. They have the ability to learn and adapt as they make decisions.

AI has benefited many areas such as chemistry and medicine, where routine diagnoses can be initiated by AI-aided computers. It embraces a wide range of disciplines such as computer science, engineering, machine learning, chemistry, biology, physics, astronomy, medicine, neuroscience, social sciences, and the military. Today, AI is integrated into our daily lives in several forms, such as personal assistants, automated mass transportation, aviation, computer gaming, facial recognition at passport control, voice recognition on virtual assistants, driverless cars, companion robots, wearables, etc. [6].

17.3 DIFFERENT TYPES OF WEARABLE DEVICES

As illustrated in Figure 17.2, AI is making wearables smarter [7].

Figure 17.2   Examples of smart wearables [7]

Wearable devices are used for many purposes such as fitness, monitoring, sleeping disorder, etc. Different types of wearable devices are discussed as follows [8],

- **Smart Phones:** People usually carry their smart phones around in their pockets. Smart phones have many sensors embedded, which collect data about the movements of users.

- **Smart Watches and Wristbands:** These have various built-in sensors that monitor users' daily activity, calorie consumption and heart rate, as well as the quality of sleep, etc. Wristbands have recently become a widely used health management tools, with large numbers of people wearing them daily.

- **Smart Glasses:** Smart glasses will not become a threat to privacy, but a practical life assistant and medical tool.

- **Smart Clothes:** Smart clothes collect body data from users, which can be used to monitor users' exercise data and heat consumption. There are also smart baby clothes for infants to monitor their physical condition.

- **Smart Shoes:** Smart sneakers collect users' sports data to help them improve their sports performance.

- **Smart Earphones:** These allow users to operate the equipment more conveniently using voice commands.

There are also smart gloves of manufacturing settings and smart suits for military and space.

17.4 AI IN WEARABLES

Wearable AI gadgets such as smartwatches and fitness bands can be used to monitor health-oriented vitals such as heart rate and blood pressure. Technologies that enable wearable AI include artificial intelligence (AI), Internet of things (IoT), and mobile devices. The Internet of things (IoT) has been driving the evolution of wearables devices, particularly in the healthcare industry. AI in wearables is aimed at improving the functionalities and user experience to provide users with real time insights. Mobile devices such as smartphones are becoming ever more important in people 's everyday lives. They have become the universal tool for accessing communications.

Typical examples of wearable-AI include [9].

- Powered by AI smart assistance and intelligence, Vinci is the wearable headphones with voice and personal assistance,

- Proteus Discover unlocks unprecedented insights into patient health patterns and medication treatment efficiencies.
- Omron Heartguide is a wearable blood pressure monitors in a wristwatch form factor that helps in tracking your heart data and learning how your behaviors impact your heart health.

- A Google Brain initiative is an AI-powered diabetic's eye disease detection.

302

- Google's AI healthcare masterpieces have been the Google smart lenses, health patch MD, Cloud DX vitality, and iTBra.

- Hearables are in-ear devices that help with fall detection by tracking activities, heart rate monitor, measuring body temperature.

- Ingestible are broadband-enabled electronic devices that are edible. "Smart" pills use wireless technology like microprocessors, power supply, sensors, etc. to help monitor internal reactions, disease diagnostics.

- Moodables provides relaxation to people suffering from stress disorders.

- Embeddable are inserted under the skin or more in-depth into the body. An example is the heart pacemaker.

17.5 APPLICATIONS OF AI IN WEARABLES

Wearable devices (WDs) are becoming widely used. Application sectors of AI-enabled wearable technologies include education, communication, navigation, entertainment, gaming, consumer electronics, business, emergency service, and military and defense, and healthcare.

Healthcare: Healthcare is transitioning from conventional medicine to a patient-centric model. AI can be a potential game-changer for healthcare by turning data into predictive knowledge for healthcare professionals. There is enormous potential in applying AI to nearly every area of healthcare. Healthcare is a unique industry that can get a huge amount of benefits from both AI and wearable technology. The integration of wearable technology in healthcare with technologies such as AI, augmented reality (AR), and virtual reality (VR) makes the wearable devices smarter. Diagnostic wearables technology can be used to diagnose various diseases in real-time. Patients can be discharged with wearable AI devices that allow remote monitoring of vitals like oxygen levels, respiratory rates, body temperature, blood pressure, and pulse [10]. An AI-based doctor may prescribe medicines for the patient. Today, AI can outperform medical practitioners in the analysis of skin lesions, pathology slides, electrocardiograms or medical imaging data. AI can help automate the EEG assessment and video segmentation process due to the large amount of data generated during this process. WDs can be used to recognize epileptic activity. There is a huge potential benefit in using WDs for seizure detection, prediction, and characterization. This could help prevent the morbidity and mortality associated with seizures. Figure 17.3 shows various uses of wearables in healthcare [11].

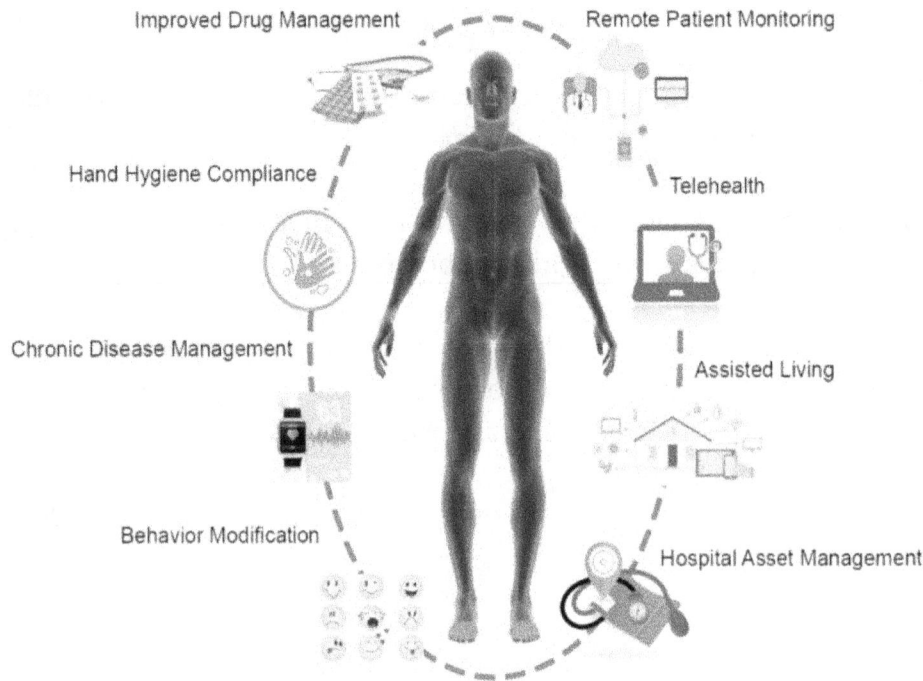

Figure 17.3 Various uses of wearables in healthcare [11].

Wearables are employed in different areas in healthcare as listed below [12].

1. Wearables for preventive health
2. Wearables for medical consultation
3. Wearables for medication management.
4. Tracking pulse and blood pressure for stress monitoring.
5. Minute-to-minute monitoring of chronic disease conditions.

Remote Monitoring: Remote monitoring of health (e.g. Parkinson disease) is a crucial application of AI wearables. Constant monitoring helps in giving adequate healthcare to patients. Patients can be monitored continuously through portable equipment. The real-time wearable monitoring technology will provide invaluable data that will interact with AI-driven video analysis and patient safety. A wireless digital watch can be used as a wearable surveillance system for monitoring the vital signs of patients [13]. Figure 17.4 illustrates wearable and flexible devices for health monitoring.

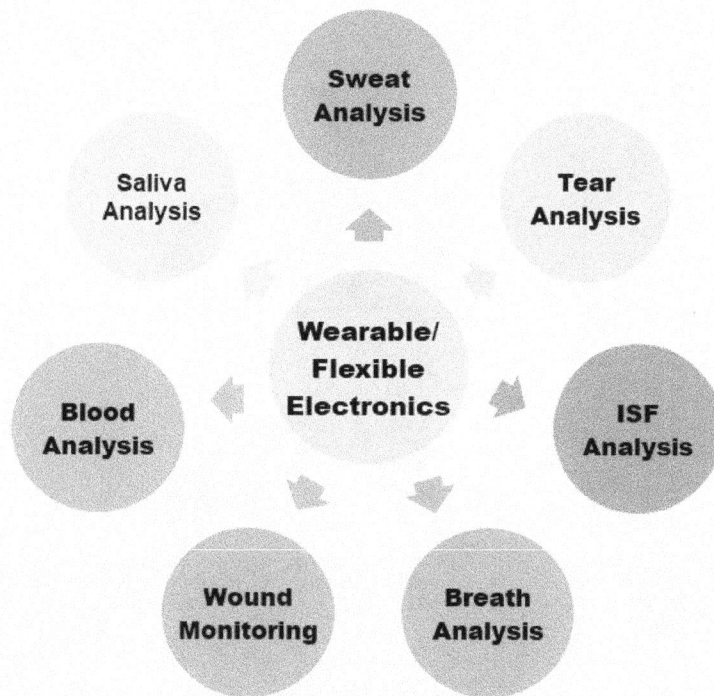

Figure 17.4 Wearable and flexible devices for health monitoring.

- **Consumer Electronics:** This sector is one of the largest demands for wearables. This is due to the growing demand of wearable devices by consumers for monitoring their routine and health-related activities. The growing need for wearables compelled companies such as Samsung Electronics, Apple, and Fitbit to focus on integrating advanced AI features in their devices. These consumer electronics companies offer advanced products at low prices, high per capita income of consumers [14].

- **Sports:** AI-based wearables are witnessing growing demand from the sports professionals. In fitness training, AI systems can provide AI app coaching in order to improve the end user's performance. AI in healthcare has enabled wearables to assist fitness and help the user to track their activities. Wearable devices have become popular for tracking physical activities and help people assess activity intensity and calories expended. Wearable devices can help athletes or coaches to manage athletic training. They can also be used as a stimulus mechanism to increase user activities. They can monitor functional movements, workloads, heart rate, etc., so they may be more widely used in sports to maximize performance and minimize injury. Sports data can be collected by smart phones, watches and wristbands, such as acceleration information, rotation speed, running steps, etc. Figure 17.5 depicts wearables in professional sports [15].

Figure 17.5 Wearables in professional sports [15].

Gaming: The gaming application segment is leveraging the rising usage of Augmented Reality (AR) and Virtual Reality (VR) in the gaming software console. The demand for gaming wearables is rising sharply as the applications developed for wearable devices have the capability to integrate motion sensing and gesture tracking to improve game experience.

Sleep Medicine: Sleep disorders affect hundreds of millions of people worldwide. Insufficient sleep increases the risk of developing serious medical conditions and shortens lifespan. Recent advances in wearable sensor technology and AI provide exciting opportunities to develop new solutions, which can monitor sleep quality and detect sleep disorders. AI can be used to predict poor sleep/sleep disorders. It will improve the practice of sleep medicine for the health and well-being of the patients [16,17].

Other areas of applications of AI-wearable include public education, enterprise, surveillance, and digital watch. Investment is rising for technological innovations to broaden the scope of AI applications of wearable devices in different end-use industries.

17.6 BENEFITS

A pair of AI-enabled spectacles has been developed to meet both fashion as well as technology needs of users. The patient can use AI-based devices to request daily walking reminders, heart rate, blood pressure, blood sugar, and weight statistics. AI wearables are making it easier for physicians and nurses to cater to the large population with real-time technology. Other benefits include:

- **Detection of Diseases:** Wearable AI gadgets can help in early detection of diseases. The data from wearable AI can be used to predict whether a person is experiencing cardiovascular disease. AI-enabled digital therapeutics and personalized recommendations will empower consumers to prevent health issues from developing. It is possible to predict health-related benchmarks during daily activities using the smart shirt.

- **Quality of Care/Life:** The wearable devices can increase the quality of care. Wearable AI system can help people navigate safely through streets and neighborhoods. Discharged patients can be fitted with a Wi-Fi-enabled armband that remotely monitors vital signs such as respiratory rate, oxygen levels, pulse, blood pressure, and body temperature. The transformation of healthcare is aimed at enhancing the quality of life for humans.

- **Monitoring:** For some conditions, it is expedient to simply monitor and record behaviors of interest. Wearable technologies offer a convenient means of monitoring many physiological features and presenting a multitude of medical solutions.

- **Safe Workplaces:** Organizations are deploying tools like AI wearable devices to protect against COVID-19 infections and ensure safe workplaces. They are also used in manufacturing settings.

17.7 CHALLENGES

Most wearable technologies are still in their prototype stages. Challenges in wearable technology still need to be addressed. There is a mismatch between what healthcare systems can do and need to deliver. Many clinicians, researchers, and decision makers are looking to AI to find the "magic bullet" to transform healthcare. More work is required to improve the accuracy and efficiency of AI-driven video analysis. Sometimes implementation of wearable devices requires users to wear a large number of devices in different parts of the body, which may cause discomfort. Video capture and editing has also been a major challenge. Other challenges include [18,19].

- **User Acceptance:** One of the barriers for wide acceptance of wearable-AI has do with ease-of-use and physical comfort. Ease-of-use is the biggest concern for individuals. User preferences should be considered in designing devices that will gain acceptance both in a clinical and home setting. For example, the devices should be well tested to meet the needs of elderly people.

- **Security of Private Information:** Patient confidentiality, privacy, and data security are major concerns when using wearable devices. There are some security sensitive AIs like wearable cameras for facial recognition. Security can be achieved by using a cryptographic scheme to ensure basic security services like confidentiality, integrity, and authenticity.

- **Privacy Concern:** This is a challenge when using wearable devices because the gathered data is stored locally at the hospital server. With the emerging success of wearable devices, the risk of threat is also increasing rapidly. A successful implementation of artificial intelligence wearables calls for keeping privacy as the top priority. Patients fear the leakage of their medical histories over wireless channels. There is a need to address issues of data transparency, privacy, and autonomy.

- **Ethical Issues:** Although advances in mobile technology offer exciting opportunities for measuring and modeling individuals' experiences in their natural environments, they also introduce new ethical issues.

- **Battery Technologies:** A common challenge in wearable-AI is the battery and energy drainage due to constant use. Conventional batteries do not meet the rising requirements of the energy storage units in wearable devices. Therefore, it is important to improve the reliability, security, and efficiency of the wearable devices through energy scavenging.

- **Fear of AI:** Many futurists have speculated about the future of artificial intelligence that could rival or exceed human intelligence. Humans should focus on teaching machines, while machines can focus on executing jobs that are too big for humans to process.

17.8 GLOBAL AI WEARABLE

The international community is currently focused on the 2019-2020 novel coronavirus (COVID-19) outbreak. As it spreads, international organizations and scientists are using AI to track the epidemic in real-time, predict where the virus might appear next, and develop an effective response. The AI wearables market is highly competitive. From fitness trackers to smartwatches, millions of users worldwide are carrying wearables. Some of the players operating in the wearable AI market include Apple, Fossil, Facebook, HTC, Bragi, Fitbit, Garmin, Jabra, Xiaomi, Huawei, Samsung, Microsoft, ANTVR, Huami, and Google. These companies produce new products that are affordable to customers worldwide and will expand the customer base.

The wearable AI market is classified based on type, application, and region. Based on type, the market is divided into smartwatches, smart glasses, smart earwear, smart gloves, etc. Based on application, the market is categorized as consumer electronics, healthcare, automotive, military & defense, media & entertainment, and etc. Based on geographical boundaries, the market is classified into North America, Europe, Asia-Pacific, and Latin America, Middle East and Africa [20]. Here we consider how AI wearable technology is implemented and used in different countries.

United States: A team of researchers at the University of Michigan recently created a wearable device that can continuously collect and examine circulating tumor cells in the blood. Amazon has launched its own fitness wearable designed to track a person's emotional well-being by scanning the tone of their voice day-to-day. Along with a weeklong battery, the water-resistant device is designed to be worn 24/7.

Canada: The wearable AI market is growing popularity of wearable technologies in Canada. The Canadian Medical Association (CMA) is providing physicians with policy guidelines in using wearable technology in mobile health applications. Other factors driving market growth are smartphone penetration, advanced connectivity infrastructure, and the rapid usage of smart wearables in the healthcare sector [21]. A team of researchers in Waterloo found that applying AI to the right combination of data retrieved from wearable technology may detect whether the condition of one's health is improving or is of concern.

Norway: This is a highly digitized nation and is at the forefront when it comes to adopting new technology. There are issues regarding transparency, the potential for abusing AI, privacy, and the "dirty data" – that is, inaccurate, manipulated or systemically biased data. These are challenges that Norway takes very seriously. The Norwegian Government recently launched a National Strategy for Artificial Intelligence that sets out how the nation will develop trustworthy AI-based technology that promotes sustainable development. Norway can become a leader in applying AI, particularly in sectors where the country already has a strong global position, such as energy, ocean industries, and health [22].

China: China is investing a great a lot in adopting wearable technology. The Chinese government has put AI at the center of its economic plan, with technology giants like Alibaba and Tencent investing billions in AI. Its goal is to become number one in the world in AI in the next decade. Given its natural advantages, like the largest smartphone market on the planet, China is likely to succeed. China had its coming-of-age AI moment with AlphaGo. The AI system, which plays the board game Go, beat the Korean champion Lee Sedol in 2016 <u>and</u> the Chinese prodigy Ke Jie earlier this year. In 2017, Chinese scientists dominate the leading AI conferences [23].

Australia: Australia's Commonwealth Scientific and Industrial Research Organization (CSIRO) developed a project called the Hospital Without Walls, which aimed to provide continuous monitoring of patients. The key technology used was a wearable, low-power radio that could transmit vital signs and activity information to a home computer. It is believed that these unobtrusive and wearable devices could advance health informatics, lead to fundamental changes of how healthcare is provided, and help to reform underfunded and overstretched healthcare systems [24].

17.9 THE FUTURE OF AI IN WEARABLES

The growing smartphone penetration coupled with development in artificial intelligence in various applications and devices is the major factor driving growth of the global wearable AI market. Today's wearables have unique features designed for increased functionality, lighter and less bulky hardware, seamless user experience, and improved connectivity. Next-generation AI wearable electronics will achieve a higher level of comfort, convenience, connection, self-sustainability, and intelligence.

We are heading toward the future where AI will help to power our human intelligence and interactions. AI and IoT are the two concepts that are expected to set new trends for wearable technology. Data and platforms, well-being and care delivery, and care enablement need to be fully integrated for the future of health to come to life. The rising prevalence of obesity and other related illnesses around the world will boost the global adoption of AI-centric wearables. The healthcare industry is on the brink of large-scale disruption. By 2040, healthcare that we know today will no longer exist. There will be a fundamental shift from "healthcare" to "health."

17.10 CONCLUSION

AI and wearables are two cutting-edge technologies that are disrupting several industries. Wearable devices and AI will improve the quality of diagnosis and patient monitoring. Wearable devices integrated with AI are becoming increasingly popular. The wearable technology industry is booming with innovations in the market at affordable prices. Wearable AI can have a significant impact on improving quality of life and well-being. Although wearable technology affords humans a multitude of capabilities unavailable to them even just a decade ago, the technology is currently in a state of infancy and there is still an opportunity to improve. For more information about artificial intelligence in wearables, one should consult the books in [25-27].

REFERENCES

[1] **M. N. O. Sadiku, P. O. Adebo, A. Ajayi-Majebi, and S.M. Musa,**" Wearable healthcare technologies," International Journal of Trend in Research and Development, vol. 7, no. 3, May-June 2020, pp. 94-97.

[2] **M. N. O. Sadiku, O. D. Olaleye, A. Ajayi-Majebi, and S. M. Musa**, "Wearable AI: A Primer," International Journal of Trend in Research and Development, vol. 8, no. 1, Jan.-Feb. 2021, pp. 35-38.

[3] "Wearable AI market size by product," https://www.gminsights.com/industry-analysis/wearable-ai-market

[4] **S. Greengard**, "What is artificial intelligence?" May 2019, https://www.datamation.com/artificial-intelligence/what-is-artificial-intelligence.html

[5] https://in.pinterest.com/pin/828662400161409072/

[6] **Y. Mintz and R. Brodie**, "Introduction to artificial intelligence in medicine," Minimally Invasive Therapy & Allied Technologies, vol. 28, no. 2, 2019, pp. 73-81.

[7] **P. Kedia**, "How to use AI to enhance today's wearables," September 2019, https://developer.qualcomm.com/blog/how-use-ai-enhance-today-s-wearables

[8] **C. Y. Jin**, "A review of AI technologies for wearable devices," IOP Conference Series: Materials Science and Engineering, 2019.

[9] "Uses and benefits of ai wearable devices for healthcare industry," https://logicsimplified.com/newgames/uses-and-benefits-of-ai-wearable-devices-for-healthcare-industry/

[10] H. Kadam, "Wearable AI and its rising penetration in healthcare industry," September 2019, |
https://www.business-newsupdate.com/wearable-ai-and-its-rising-penetration-in-healthcare-industry.

[11] https://www.researchgate.net/figure/Illustration-of-an-ambient-assisted-living-system_fig4_322261039

[12] "Implementing AI in wearable health apps for better tomorrow,"
https://www.google.com/search?q=Implementing+AI+in+wearable+health+apps+for+better+tomorrow&rlz=1C1CHBF_enUS910US910&oq=Implementing+AI+in+wearable+health+apps+for+better+tomorrow&aqs=chrome..69i57j69i60.2627j0j7&sourceid=chrome&ie=UTF-8

[13] **Y. Yang and W. Gao**, "Wearable and flexible electronics for continuous molecular monitoring," April 2018.
https://www.researchgate.net/figure/Wearable-and-flexible-chemical-sensors-for-non-invasive-health-monitoring_fig1_324206200

[14] "Research and markets releases report: Wearable AI devices market research report," February 2020,

https://www.researchandmarkets.com/reports/4912441/wearable-ai-devices-market-research-report-
by?utm_source=dynamic&utm_medium=BW&utm_code=4krqq9&utm_campaign=1347241+-
+Global+Wearable+AI+Devices+Market+Industry+Predicted+to+Grow+with+a+CAGR+of+29.0%25+by+the+End+of+2024&utm_exec=anwr281bwd

[15] **E. Waltz**, "Rocky start for wearables in professional sports games,"
https://spectrum.ieee.org/the-human-os/biomedical/devices/rocky-start-to-wearables-in-professional-sports

[16] "Machine learning and wearable technology in sleep medicine,"
https://www.frontiersin.org/research-topics/16391/machine-learning-and-wearable-technology-in-sleep-medicine

[17] **C. A. Goldstein** et al., "Artificial intelligence in sleep medicine: Background and implications for clinicians, "Journal of Clinical Sleep Medicine, vol. 16, no. 4, April 2020, pp. 609-618.

[18] **M. Wu and J. Luo**, "Wearable technology applications in healthcare: A literature review," Online Journal of Nursing Informatics, vol. 23, no. 3, Fall 2019.

[19] **M. M. El Khatib and G. Ahmed**, "Management of artificial intelligence enabled smart wearable devices for early diagnosis and continuous monitoring of CVDS," International Journal of Innovative Technology and Exploring Engineering, vol. 9, no, 1, November 2019, pp. 1211-1215.

[20] **S. Bansal**, "Wearable AI- Implementation and benefits in digital world," January 2021,
https://www.experfy.com/blog/ai-ml/wearable-ai-implementation-benefits-digital-world/

[21] "Wearable AI Market Size By Product (Smartwatch & Fitness Band, Head Mounted Display (HMD), Earwear), By Application (Consumer Electronics, Gaming, Enterprise), Industry Analysis Report, Regional Outlook, Growth Potential, Competitive Market Share & Forecast, 2019 – 2025," April 2019,
https://www.gminsights.com/industry-analysis/wearable-ai-market

[22] **J. Sortino**, "This is how Norway puts artificial intelligence to use," February 2020,

https://www.theexplorer.no/stories/technology/this-is-how-norway-puts-artificial-intelligence-to-use/#:~:text=The%20Norwegian%20company%20Scantrol%20Deep,the%20catch%20onboard%20the%20vessel.

[23] **S. Simone**, "Artificial intelligence, according to these 5 Experts," November 2017, https://www.delltechnologies.com/en-us/perspectives/artificial-intelligence-according-to-these-5-experts/

[24] **M. Wu and J. Luo**, "Wearable technology applications in healthcare: A literature review," Online Journal of Nursing Informatics, vol. 23, no. 3, Fall 2019.

[25] **J. L. Pons** (ed.), Wearable Robots: Biomechatronic Exoskeletons. John Wiley & Sons, 2008.

[26] **Y. Xu, W. J. Li, and K. K. Lee**, Intelligent Wearable Interfaces. Wiley-Interscience, 2010.

[27] **E. Sazonov** (ed.), Wearable Sensors: Fundamentals, Implementation And Applications. Academic Press, 2014.

CHAPTER 18
AI IN CYBERSECURITY

"It takes 20 years to build a reputation and few minutes of cyber-incident to ruin it."
– *Stephane Nappo*

18.1 INTRODUCTION

Today, we live in cyber world where everything is digital, and data is king. Data security is now more important than ever. Hackers are becoming smarter day by day and they are more innovative in exploiting the vulnerable data of individuals, organizations, and governments. New cyberattacks, data bridges, and data breaches, data poisoning, hackers attacks, and crashes come to light almost every day. Cyber criminals pose a threat to organizations, businesses, governments, and consumers who use computer networks. Cyber-attacks have been ranked as one of the top five most likely sources of severe, global-scale risk. Attacks to networks are becoming more complex and cybercriminals are becoming more sophisticated every day. This compels organizations and other users of computer networks to pay close attention to their network security.

Cybersecurity is the process of protecting computer networks from cyber-attacks or unintended, unauthorized access. It is the need of the hour. Organizations, businesses, and governments need cybersecurity solutions because cyber criminals pose a threat to everyone. AI promises to be a great solution for this. By combining the strength of artificial intelligence with cybersecurity, security experts are more capable to defend vulnerable networks and data from cyber attackers [1].

This chapter provides an introduction to the use of artificial intelligence in cybersecurity. It begins by briefly reviewing AI. It explains the concept of cybersecurity. It covers some applications of AI in cybersecurity. It highlights the benefits and challenges of AI in cybersecurity. It describes the global applications of AI in cybersecurity. It discusses the future of AI in cybersecurity.   The last section concludes with comments.

18.2 REVIEW ON ARTIFICIAL INTELLIGENCE

Artificial intelligence (AI) refers to computer systems that mimic human cognitive functions. It is a field of computer science that deals with intelligent machines. AI has long history which is actively and constantly changing and growing. The term "artificial intelligence" was first used at a Dartmouth College conference in 1956. The main goal of AI is to enable machines to perform complex tasks that typically require human intelligence [2]. In simple terms, AI attempts to clone human behavior. An important feature of AI technology is that is can be added to existing technologies. AI is now one of the most important global issues of the 21st century. It is poised to disrupt our world and change processes and developments in the fields of science, engineering, education, business, entertainment, and agriculture.

Artificial intelligence is an umbrella term that encompasses many different technologies. AI is not a single technology but a collection of techniques that enables computer systems to perform tasks that would otherwise require human intelligence [3]. The major disciplines in AI include:

- Expert systems
- Fuzzy logic
- Neural networks
- Machine learning (ML)
- Deep learning
- Natural Language Processors (NLP)
- Robots
- Data Mining:

These AI tools are illustrated in Figure 18.1.

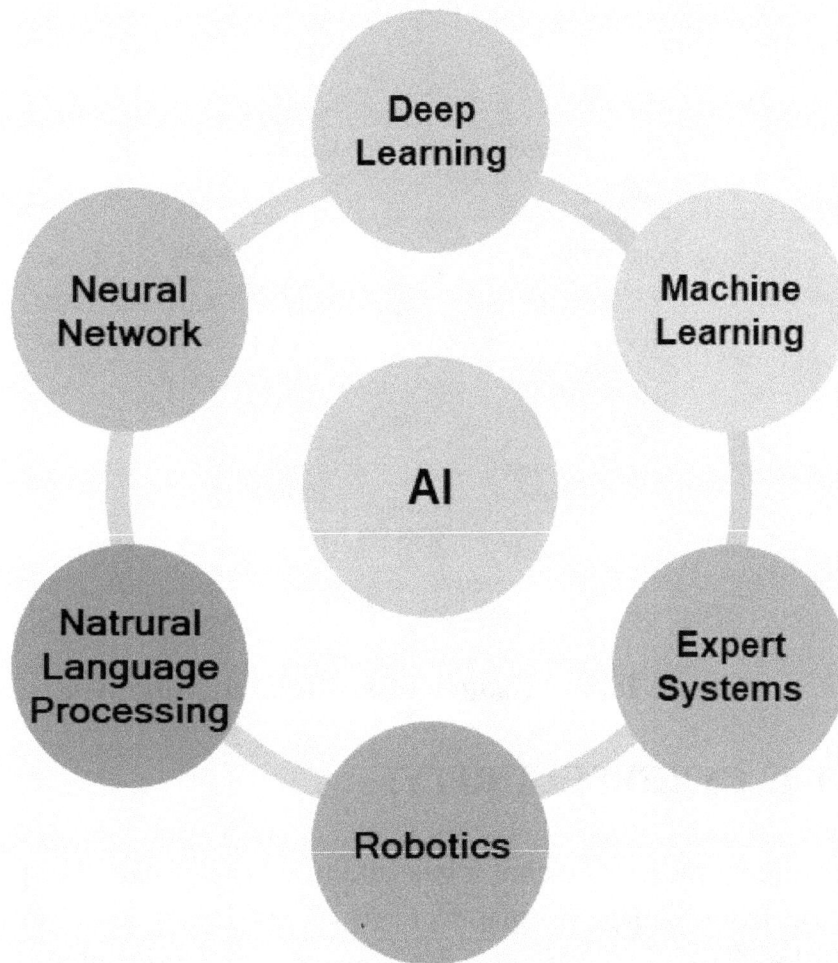

Figure 18.1 Different branches of artificial intelligence.

Each AI tool has its own advantages. Using a combination of these models, rather than a single model, is recommended. AI systems are designed to make decisions using real-time data. They have the ability to learn and adapt as they make decisions.

AI has benefited many areas such as chemistry and medicine, where routine diagnoses can be initiated by AI-aided computers. It embraces a wide range of disciplines such as computer science, engineering, chemistry, biology, physics, astronomy, medicine, neuroscience, social sciences, and the military. Today, AI is integrated into our daily lives in several forms, such as personal assistants, automated mass transportation, aviation, computer gaming, facial recognition at passport control, voice recognition on virtual assistants, driverless cars, companion robots, wearables, etc. [4,5]. Figure 18.2 illustrates some usages of artificial intelligence.

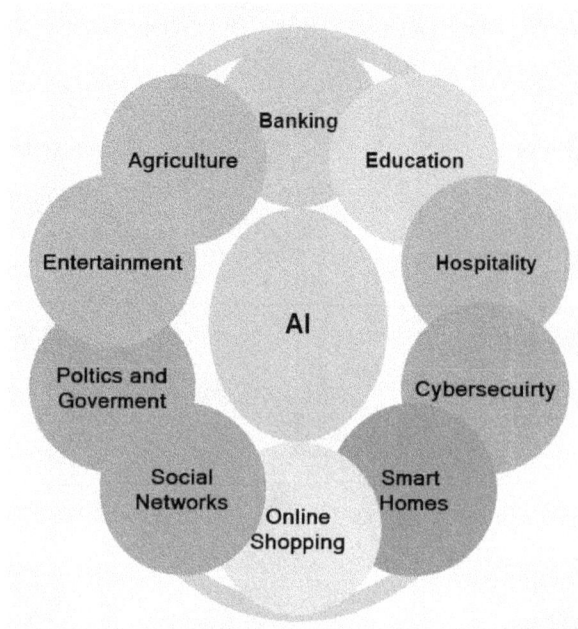

Figure 18.2   Usages of artificial intelligence.

18.3 WHAT IS CYBERSECURITY?

Cybersecurity refers to a set of technologies and practices designed to protect networks and information from damage or unauthorized access. It is vital because governments, companies, and military organizations collect, process, and store a lot of data. As shown in Figure 18.3, cybersecurity involves multiple issues related to people, process, and technology.

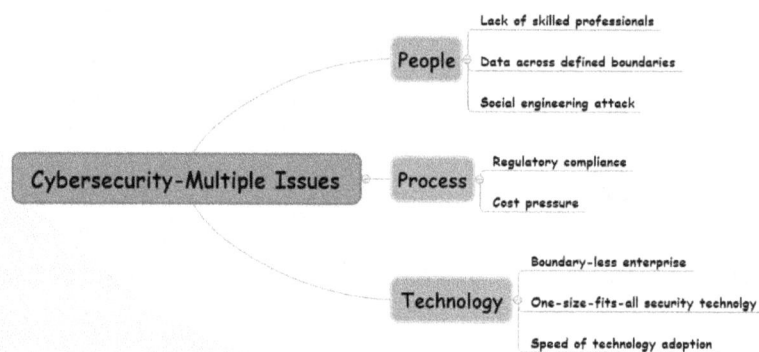

Figure 18.3   Cybersecurity involves multiple issues related to people, process, and technology.

A typical cyber-attack is an attempt by adversaries or cybercriminals to gain access to and modify their target's computer system or network. Cyber-attacks are becoming more frequent, sophisticated, dangerous, and destructive. They are threatening the operation of businesses, banks, companies, and government networks. They vary from illegal crime of individual citizen (hacking) to actions of groups (terrorists) [6,7].

The cybersecurity is a dynamic, interdisciplinary field involving information systems, computer science, and criminology. The security objectives have been availability, authentication, confidentiality, nonrepudiation, and integrity. A security incident is an act that threatens the confidentiality, integrity, or availability of information assets and systems [8].

- **Availability:** This refers to availability of information and ensuring that authorized parties can access the information when needed. Attacks targeting availability of service generally leads to denial of service.

- **Authenticity:** This ensures that the identity of an individual user or system is the identity claimed. This usually involves using username and password to validate the identity of the user. It may also take the form of what you have such as a driver's license, an RSA token, or a smart card.

- **Integrity:** Data integrity means information is authentic and complete. This assures that data, devices, and processes are free from tampering. Data should be free from injection, deletion, or corruption. When integrity is targeted, nonrepudiation is also affected.

- **Confidentiality:** Confidentiality ensures that measures are taken to prevent sensitive information from reaching the wrong persons. Data secrecy is important especially for privacy-sensitive data such as user personal information and meter readings.

- **Nonrepudiation:** This is an assurance of the responsibility to an action. The source should not be able to deny having sent a message, while the destination should not deny having received it. This security objective is essential for accountability and liability.

Everybody is at risk for a cyber-attack. Cyber-attacks vary from illegal crime of individual citizen (hacking) to actions of groups (terrorists). The following are typical examples of cyber-attacks or threats [9]:

- **Malware:** This is a malicious software or code that includes traditional computer viruses, computer worms, and Trojan horse programs. Malware can infiltrate your network through the Internet, downloads, attachments, email, social media, and other platforms. Spyware is a type of malware that collects information without the victim's knowledge.

- **Phishing:** Criminals trick victims into handing over their personal information such as online passwords, social security numbers, and credit card numbers.

- **Denial-of-Service Attacks:** These are designed to make a network resource unavailable to its intended users. These can prevent the user from accessing email, websites, online accounts or other services.

- **Social Engineering Attacks:** A cyber-criminal attempts to trick users to disclose sensitive information. A social engineer aims to convince a user through impersonation to disclose secrets such as passwords, card numbers, or social security numbers.

- **Man-In-the-Middle Attack:** This is a cyber-attack where a malicious attacker secretly inserts him/herself into a conversation between two parties who believe they are directly communicating with each other. A common example of man-in-the-middle attacks is eavesdropping. The goal of such an attack is to steal personal information.

Cybersecurity involves reducing the risk of cyber-attacks. Cyber risks should be managed proactively by the management. Cybersecurity technologies such as firewalls are widely available [10]. Cybersecurity is the joint responsibility of all relevant stakeholders including government, business, infrastructure owners, and users. Cybersecurity experts have shown that passwords are highly vulnerable to cyber threats, compromising personal data, credit card records, and even social security numbers. Governments and international organizations play a key role in cybersecurity issues. Securing the cyberspace is of high priority to the US Department of Homeland Security (DHS). Vendors that offer mobile security solutions include Zimperium, MobileIron Skycure, Lookout, and Wandera.

18.4 APPLICATIONS OF AI IN CYBERSECURITY

There is a wide range of interdisciplinary intersections between AI and cybersecurity. AI tools (such as expert systems, computational intelligence, neural networks, intelligent agents, artificial immune systems, machine learning, data mining, pattern recognition, fuzzy logic, heuristics, etc.) have been increasingly applied for cyber-crime detection and prevention [11]. They can be used to learn how to enable security experts to understand the cyber environment in order to detect abnormalities. The use of AI can help broaden the horizons of existing cyber security solutions. Figure 18.4 shows some applications of AI for cybersecurity.

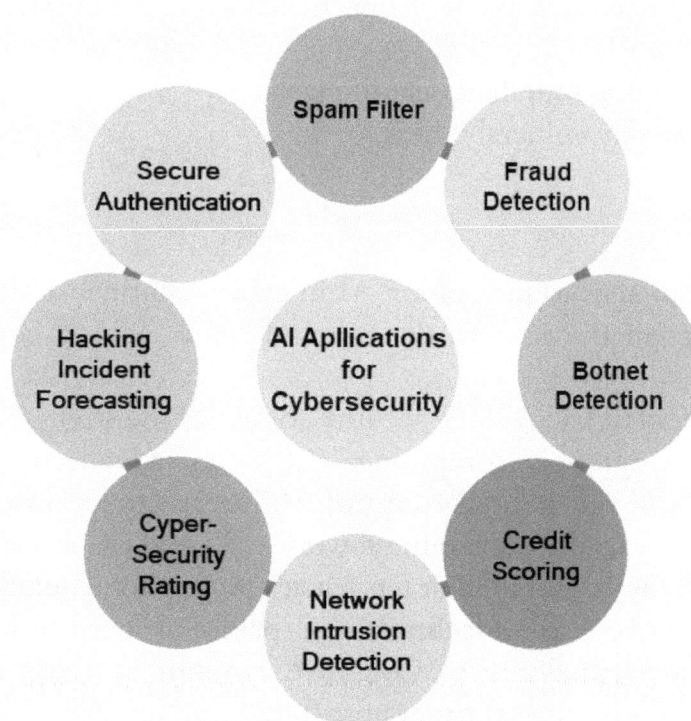

Figure 18.4 Applications of AI for cybersecurity.

To enhance existing cybersecurity systems, companies can apply AI in the following four areas: automated defense, cognitive security, adversarial training, parallel and dynamic monitoring.

- **Automated Defense:** There are two types of cybersecurity systems: expert (analyst-driven) and automated (machine-driven). Expert systems are developed and operated by people, while automated systems use AI intelligent tools. AI-based systems are autonomous, self-learning agents. A good example of automated system is CAPTCHA (Completely Automated Public Turing test to tell Computers and Humans Apart). The speed and amount of data necessary to defend the cyber space cannot be handled by humans without automation. As networks become larger and more complex, AI can be a huge boon to an organization's cyber protections. An ideal cyber-defense aims at fully protecting users, while preserving all the functionalities. AI automated systems can be integrated into existing cyber security functions. Some of these functions include [12-14]:

- Creating more accurate, biometric-based login techniques
- Detecting threats and malicious activities using predictive analytics
- Enhancing learning and analysis through natural language processing
- Securing conditional authentication and access
- Improving human analysis – from malicious attack detection to endpoint protection
- Using in automating mundane security tasks
- Having no zero-day vulnerabilities

There are many advantages to integrating AI in cyber security. AI-based cybersecurity solutions are designed to work around the clock to protect you.

- **Cognitive Security:** Cognitive security combines the strengths of AI and human intelligence. Cognitive computing (CC), an advanced type of artificial intelligences, leverages various forms of AI. It refers to hardware and/or software that mimics the way the human brain works. AI and CC remain closely similar in the intent, but they differ in their tendencies to interact naturally with humans. AI has been described as a technology capable of performing tasks normally requiring human intelligence. Cognitive computing seeks to overcome the boundaries of conventional programmable (von Neumann) computers. Watson for cybersecurity, IBM's first cognitive system, demonstrated through a Jeopardy exhibition match that it was capable of answering complex questions as effectively as the world's human champions. Watson learns with each interaction to connect the dots between threats and provide actionable insights. This allows an analyst to respond to threats with greater confidence and speed [15].

- **Adversarial Training:** This term is often used to refer to the development and use of AI for malicious purposes. Cybersecurity engineers are creating pre-emptive adversarial attack models to probe AI vulnerabilities. An adversarial learning attack can cause the algorithms to misbehave or reveal information about their inner workings.

324

- Adversarial training between AI systems can help to improve their robustness as well as facilitate the identification of vulnerabilities of the system. The more we adopt an "adversarial" and "always-on" strategy, the safer the AI applications become [16].

- **Parallel and Dynamic Monitoring:** The learning abilities of the targeted systems require some form of constant monitoring during deployment. Monitoring is necessary to ensure that divergence between the expected and actual behavior of a system is captured and addressed adequately. To do so, providers of AI systems should maintain a clone system as a control system, which serves as the benchmark against which the behavior of the original system is assessed [16].

18.5 BENEFITS

Artificial intelligence is playing a key role in digital transformation through its automated decision-making capabilities. It is well suited to solve some of the most difficult problems including cybersecurity. It is an enabling technology that transforms our lives. Embedded in our homes, cars, and devices, it will make everything "smarter" and more efficient. New AI capabilities can make the world safer, just, and environmentally friendly. Due to their flexibility and adaptive behavior, AI-based techniques can help us overcome the deficiencies of conventional cybersecurity tools. It is capable of adapting and responding to a constantly changing world. It improves how cybersecurity experts analyze, study, and understand cyber-crimes. A key benefit of machine learning in cybersecurity is that it identifies and promptly reacts to suspected problems. Other benefits include [17]:

- **Protection:** Cybersecurity protects citizens and society from harm and attacks perpetrated through computer networks. Cybersecurity helps companies better address some cybersecurity challenges such as the impact of a cyberattack, huge financial loss, business disruption, brand reputation damage, customers' privacy, data protection, etc.

- **Smart Security:** Cyber attackers are getting smarter. Your security should be getting smarter too. Businesses now depend on AI technologies, like machine learning, deep learning, and natural language processing, to help security analysts to respond to threat with speed and accuracy. They also help protect networks and sensitive data.

- **Accurate prediction:** AI can be used to predict security breaches or cyber threats.

- **Faster response:** AI proves its efficiency in helping organizations to respond faster to the next generation of cyberattacks. AI tools for cybersecurity have helped to reduce breach risk and improve their security posture efficiently and effectively.

- **Detection of Threats:** AI can be used to detect threats and other malicious activities. The ability of AI to analyze massive data with lightning speed implies that security threats can be detected in real time, or even predicted based on risk modeling. AI can detect whether a software is a malware or a normal software.

18.6 CHALLENGES

Although AI tools could help fight cybercrime, the tools are not a silver bullet and could be exploited by malicious hackers. Some regard AI in cybersecurity as posing both a blessing and a curse. AI technology can be used for good and bad, although the good outweighs the bad. There are limitations which prevent AI from becoming a mainstream tool. Cybersecurity is a domain where absolute security is impossible.

If a machine learning-based security tool misses a particular kind of cyberattack because it is not coded into it, that may lead to problems. Hackers themselves can use AI to test and develop their malware and make it potentially become AI-proof. Besides, shortage of cybersecurity experts is another problem. Other challenges include the following:

- **Potential Downsides:** The downsides of AI in cyber security include cost, intensive resources, and training. AI in cybersecurity requires more resources and finances than traditional non-AI cyber security solutions and may not be practical in some applications.

- **Use by Adversaries:** Just as AI technologies can be used to identify and stop cyberattacks, cybercriminals can also use the AI systems to launch counter attacks, defeat defenses, and avoid detection. Then AI must be developed to devise counterattacks.

- **Massive Data:** It is needless to say that AI and massive data go hand in hand. The massive amounts of data have moved beyond a human-scale problem.

- **AI technology:** AI tools themselves bring some security issues that need to be solved. For example, data mining and machine learning create a wealth of privacy issues due to the abundance and accessibility of data.

- **Cybersecurity Myths:** AI tools can get you into trouble if you depend on them as an oracle of everything. The tools are not a silver bullet. They will not solve all your problems. Some critics have warned that AI could make cyberattacks more dangerous and more difficult to spot than ever before [18].

These challenges prevent AI from becoming the only cybersecurity solution.

18.7 GLOBAL AI IN CYBERSECURITY

The cyberspace may be regarded as a domain of warfare, and AI is a new defence capability. In 2019, the World Economic Forum ranked cyber-attacks among the top-ten most impactful global risks, indicating that cyber-attacks pose an ever-growing threat to information community. Research on AI for cyber defence is progressing quickly with the United States leading technologically.

The Global AI in cybersecurity market is segmented based on security type, technology, application, and geography. Based on security type, the market is classified into Application Security, Cloud Security, Endpoint Security, and Network Security. Based on technology, the market is bifurcated into Context-Aware Computing, Machine Learning, and Natural Language Processing. Based on application, the market is divided into Antivirus/Antimalware, Data Loss Prevention, Fraud Detection/Anti-Fraud, Intrusion Detection/Prevention System, Security & Vulnerability Management, Unified Threat Management, etc. Based on geography, the market is classified into North America, Europe, Asia Pacific, and Rest of the world. The key players in the global market include Intel, IBM, Amazon Web Services, Samsung Electronics, NVIDIA Corporation, and Vectra AI [19]. We now consider how AI in cybersecurity is implemented around the world.

- **United States:** As individuals, corporations, and governments generate increasing amounts of data, they face new cyber-threats against networks, information systems and infrastructures. The US Department of Homeland Security (DIH) has developed a system called AVATAR that leverages AI and big data to pick up body gestures and facial expressions of people. DHS has a dedicated division responsible for risk management program and requirements for cybersecurity called the National Cyber Security Division. The Federal Communications Commission's role in cybersecurity is to strengthen the protection of critical computer networks and networked infrastructure. The Computer Fraud and Abuse Act (CFAA) remains the most relevant applicable law expressing the US proactive cybersecurity effort [20].

  - **United Kingdom:** In 2017, the software firm DarkTrace in Cambridge launched Antigena, which uses machine learning to spot abnormal behavior on an IT network, shut down communications to that part of the system, and issue an alert. The National Cyber Security Centre (NCSC) launched an Active Cyber Defence Programme, which fosters forms of network monitoring to identify attacks and sources of attacks and enables some forms of threat response.

- **China:** Artificial intelligence is a priority for China, which aims to become a world leader in machine-learning technologies. In 2017, the Chinese government issued its next generation AI Development Plan. Military implementation of AI, on the battlefield as well as in cyberspace, is a crucial part of the strategy. China has begun employing AI to attempt the prediction of crimes before. By using AI and facial recognition to identify people, Chinese authorities can gather information on people and their activities.

- **European Union:** Cybersecurity will serve as a springboard for the widespread secure deployment of AI across the EU. In 2017, EU reassessed cybersecurity and defense policies and launched the European Centre of Excellence for Countering Hybrid Threats, based in Helsinki. The EU has the most comprehensive regulatory framework for state conduct in cyberspace. The EU treats cyberdefence as a case of cybersecurity, disregards active uses of cyberdefence, and does not include AI. The EU must take the following three steps to avoid serious imminent attacks on state infrastructures, and to maintain international stability: (1) Define legal boundaries; (2) Test strategies with allies; (3) Monitor and enforce rules. The EU aims at developing the "AI made in Europe" brand as a seal of quality for ethical, secure and cutting-edge AI [21].

- **Russia:** The nation has not released any public documents about its strategies for AI in defense. However, in 2017, President Vladimir Putin referred to AI and stated: "Whoever becomes the leader in this sphere will become the ruler of the world." Experts agree that Russia is focusing on developing AI-enhanced tools for its conventional forces. Since 2014, the Russian National Defense Control Center has been using machine-learning algorithms to detect online threats [22].

18.8 FUTURE OF AI IN CYBERSECURITY

Cybersecurity is undergoing massive shifts in technology and its operations in recent days. It faces a number of challenges such as intrusion detection, privacy protection, proactive defense, anomalous behaviors, advanced threat detection, etc. The majority of cybersecurity traditional tools require human interaction at some point. The adoption of AI in cybersecurity has changed the entire scenario. In the coming years, companies are likely to rely on AI tools to prevent cyber-attacks.

The era of AI is around the corner. The future of AI-enabled cybersecurity is very promising. Artificial Intelligence, which is driving a revolution in almost every industry, can be the catalyst in increasing the effectiveness of cybersecurity. The application of AI to cybersecurity is an emerging field. AI is the future for cybersecurity. We can expect to apply AI in a whole new generation of products to improve safety and security and reduce the necessary costs for its management. For future AI and cybersecurity technology to be successful, the quality data needs to be improved.

18.9 CONCLUSION

Artificial intelligence has become a growing area of interest and investment within the cybersecurity community. It is, without question, one of the industries of the future. Some early AI adopters include Google, IBM, Juniper Networks, Apple, Amazon, and Balbix. An increasing number of companies and organizations are jumping on the AI bandwagon. The emergence of AI improves cybersecurity technologies and is used to take action against cybercriminals.

Cybersecurity has become a major concern in the digital era. It has always been an arms race. It is one of the biggest challenges of the digital age and it keeps getting bigger. Security incidents such as data breaches, identity theft, unauthorized access, malware attack, denial of service, etc. abound and affect millions of individuals and organizations. As cyber-attacks and crimes grow, artificial intelligence is helping security operations analysts come to grips with and even stay ahead of threats. AI automated systems will soon become an integral part of cybersecurity solutions, but it will also be used by cybercriminals to do harm. More information on artificial intelligence in cybersecurity can be found in the books in [12,23-33] and the following related journals:

- Journal of Computer Security

- Artificial Intelligence Review

- International Journal of Artificial Intelligence & Applications

- International Journal of Computer and Internet Security

- International Journal of Cyber Criminology

REFERENCES

[1] **M. N. O. Sadiku, O. I. Fagbohungbe**, and **S. M. Musa**, "Artificial intelligence in cyber security," International Journal of Engineering Research and Advanced Technology, vol. 6, no. 5, 2020, pp.1-8.

[2] **S. Greengard**, "What is artificial intelligence?" May 2019, https://www.datamation.com/artificial-intelligence/what-is-artificial-intelligence.html

[3] https://in.pinterest.com/pin/828662400161409072/

[4] "Driving business results with artificial intelligence services," September 2020, https://laptrinhx.com/driving-business-results-with-artificial-intelligence-services-661985586/

[5] **Y. Mintz** and **R. Brodie**, "Introduction to artificial intelligence in medicine," Minimally Invasive Therapy & Allied Technologies, vol. 28, no. 2, 2019, pp. 73-81.

[6] **A. M. Alrajhi**, "A survey of artificial intelligence techniques for cybersecurity improvement," International Journal of Cyber-Security and Digital Forensics, vol. 9, no.1, pp. 34-41.

[7] "Eliminating the complexity in cybersecurity with artificial intelligence," https://www.wipro.com/cybersecurity/eliminating-the-complexity-in-cybersecurity-with-artificial-intelligence/

[8] **M. N. O. Sadiku, M. Tembely**, and **S. M. Musa**, "Smart grid cybersecurity," Journal of Multidisciplinary Engineering Science and Technology, vol. 3, no. 9, September 2016, pp.5574-5576.

[9] FCC Small Biz Cyber Planning Guide, https://transition.fcc.gov/cyber/cyberplanner.pdf

[10] **Y. Zhang**, "Cybersecurity and reliability of electric power grids in an interdependent cyber-physical environment," Doctoral Dissertation, University of Toledo, 2015.

[11] **S. Dilek, H. Çakır**, and **M. Aydın**, "Applications of artificial intelligence techniques to combating cybercrimes: A review," International Journal of Artificial Intelligence & Applications, vol. 6, no. 1, January 2015, pp. 21-39.

[12] **R. Prasad** and **V. Rohokale**, Cyber Security: The Lifeline of Information and Communication Technology. Springer, 2020.

[13] **C. Crane**, "Artificial intelligence in cyber security: The savior or enemy of your business?" July 2019, https://www.thesslstore.com/blog/artificial-intelligence-in-cyber-security-the-savior-or-enemy-of-your-business/

[14] "The role of AI in cybersecurity," https://blog.eccouncil.org/the-role-of-ai-in-cybersecurity/

[15] "Artificial intelligence for a smarter kind of cybersecurity," https://www.ibm.com/security/artificial-intelligence

[16] **M. Taddeo, T. McCutcheon**, and **L. Floridi** , "Trusting artificial intelligence in cybersecurity is a double-edged sword," Nature Machine Intelligence, vol. 1, November 2019, pp. 557–560.

[17] "How artificial intelligence helps companies better focus on cybersecurity," https://def.camp/artificial-intelligence-cybersecurity/

[18] **D. Palmer,** "AI is changing everything about cybersecurity, for better and for worse. Here's what you need to know," March 2020, https://www.zdnet.com/article/ai-is-changing-everything-about-cybersecurity-for-better-and-for-worse-heres-what-you-need-to-know/

[19] "Artificial intelligence in cyber security market size and forecast," https://www.verifiedmarketresearch.com/product/artificial-intelligence-in-cyber-security-market/

[20] **M. N. O. Sadiku, S. Alam, S. M. Musa,** and **C. M. Akujuobi,**" A primer on cybersecurity," International Journal of Advances in Scientific Research and Engineering, vol. 3, no. 8, Sept. 2017, pp. 71-74.

[21] European Union Agency for Cybersecurity (ENISA), "AI cybersecurity challenges: Threat landscape for artificial intelligence," December 2020.

[22] **I. H. Sarkar et al.,** "Cybersecurity data science: An overview from machine learning perspective," Journal of Big Data, vol 7, no. 41, July 2020.

[23] **N. J. Daras** and **M. T. Rassias (eds.),** Computation, Cryptography, and Network Security. Springer, 2015.

[24] **A. Parisi,** Hands-On Artificial Intelligence for Cybersecurity: Implement Smart AI Systems for Preventing Cyber Attacks and Detecting Threats and Network Anomalies. Packt Publishing, 2019.

[25] **S. Halder** and **S. Ozdemir,** Hands-On Machine Learning for Cybersecurity: Safeguard your system by making your machines intelligent using the Python ecosystem. Packt Publishing, 2018.

[26] **M. Gilbert (ed.),** Artificial Intelligence for Autonomous Networks. Boca Raton, FL: CRC Press, 2018.

[27] **B. B. Gupta** and **Q. Z. Sheng,** Machine Learning for Computer and Cyber Security: Principle, Algorithms, and Practices. Boca Raton, FL: CRC Press, 2019.

[28] **L. F. Sikos (ed.),** AI in Cybersecurity. Springer, 2019.

[29] **T. Coombs**, Artificial Intelligence & Cybersecurity for Dummies®, IBM Limited Edition. Hoboken, NJ: John Wiley & Sons, 2018.

[30] **D. Ventre**, Artificial Intelligence, Cybersecurity and Cyber Defence. John Wiley & Sons, 2020.

[31] **F. Liu et al.,** Science of Cyber Security. Springer, 2018.

[32] National Academies of Sciences, Engineering, and Medicine, Implications of Artificial Intelligence for Cybersecurity: Proceedings of a Workshop. The National Academies Press, 2019.

[33] **Y. R. Masakowski**, Artificial Intelligence and Global Security: Future Trends, Threats and Considerations. Emerald Publishing Limited, 2020.

CHAPTER 19
AI IN MILITARY

"Artificial intelligence is the future, not only for Russian, but for all of humankind. It comes with colossal opportunities, but also threats that are difficult to predict. Whoever becomes the leader in this sphere will become the ruler of the world." ~ *Vladimir Putin.*

19.1 INTRODUCTION

Technological development has become a rat race. New technologies that promise significant strategic advantages can upset balances or disrupt previously stable global governance arrangements. Artificial intelligence (AI) is one such critical technology. AI is an integral part of bringing technological advancements to the next level. It is among the many hot technologies that promise to change the face of warfare for years to come [1,2].

Artificial intelligence (AI) has done remarkable things such as defeating human experts at various games. AI is a technology that the military and defense world cannot ignore because the military cannot afford to miss out on the opportunities it brings. It has been described as the "third revolution" in warfare, after gunpowder and nuclear weapons. It has also been considered as the "fourth industrial revolution," which includes the Internet of things (IoT), nanotechnology, biotechnology, and robotics. It has become a critical part of modern warfare. It could cause drastic changes in hybrid warfare, which is a major concern for NATO [3].

Advances in AI, machine learning, and robotics are enabling new military capabilities that will have a disruptive impact on military strategies. The effects of these capabilities will be felt across the spectrum of military requirements – from intelligence, surveillance, marketing departments, and reconnaissance to offense/defense balances and even on to nuclear weapons systems themselves. Artificial intelligence and other emerging technologies will change the way war is fought. Whether it involves AI or not, war will always be violent, politically motivated, and composed of the same three elemental functions that new recruits learn in basic training: move, shoot, and communicate [4].

This chapter examines various applications of artificial intelligence in the military and defense. It begins by briefly reviewing the concept of AI. It covers what military is all about. It presents how AI is being incorporated in the military.

It covers some applications of AI in the military. It highlights the benefits and challenges of military AI. It addresses how AI is being integrated in military forces around the globe. It presents the future of military AI. The last section concludes with comments.

19.2 REVIEW ON ARTIFICIAL INTELLIGENCE

Artificial intelligence (AI) refers to computer systems that mimic human cognitive functions. It is a field of computer science that deals with intelligent machines. AI has long history which is actively and constantly changing and growing. The term "artificial intelligence" was first used at a Dartmouth College conference in 1956. The main goal of AI is to enable machines to perform complex tasks that typically require human intelligence [5]. In simple terms, AI attempts to clone human behavior. An important feature of AI technology is that is can be added to existing technologies. AI is now one of the most important global issues of the 21st century. It is poised to disrupt our world and change processes and developments in the fields of science, engineering, education, business, entertainment, agriculture, and military.

Artificial intelligence is an umbrella term that encompasses many different technologies. AI is not a single technology but a collection of techniques that enables computer systems to perform tasks that would otherwise require human intelligence [6]. The major disciplines in AI include:

- Expert systems
- Fuzzy logic
- Neural networks
- Machine learning (ML)
- Deep learning
- Natural Language Processors (NLP)
- Robots

These AI tools are illustrated in Figure 19.1.

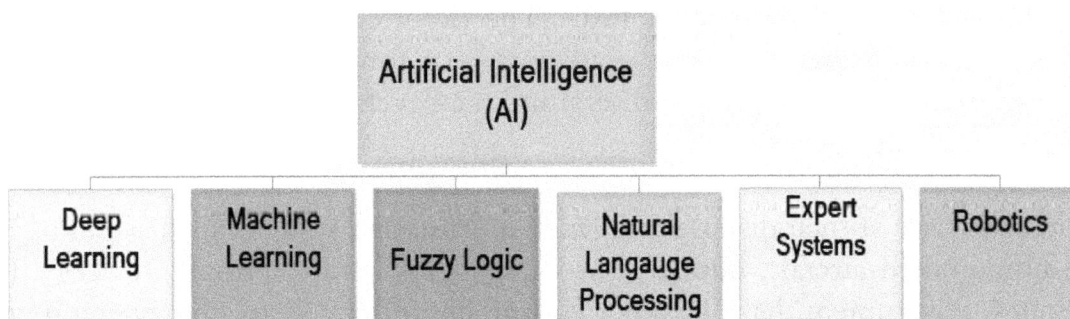

Figure 19.1 Branches of artificial intelligence.

Each AI tool has its own advantages. Using a combination of these models, rather than a single model, is recommended. AI systems are designed to make decisions using real-time data. They have the ability to learn and adapt as they make decisions.

AI has benefited many areas such as chemistry and medicine, where routine diagnoses can be initiated by AI-aided computers. Today, AI is integrated into our daily lives in several forms, such as personal assistants, automated mass transportation, aviation, computer gaming, facial recognition at passport control, voice recognition on virtual assistants, driverless cars, companion robots, wearables, etc. AI is emerging as the base technology for the military systems, where the future intelligent weapons are envisaged to transform the military operations. Now AI technologies are widely used in tactical warfare situations, such as target acquisition for missiles launched from drones. Military robots are usually employed within integrated systems that include video screens, sensors, grippers, and cameras [7,8].

19.3 MILITARY

The military, also known collectively as armed forces, is an armed and organized force primarily intended for warfare. It may consist of several branches such as army, navy, air force, space force, marines, and coast guard. The major job of the military is defined as defense of the state and its interests against external armed threats [9].

The US Department of Defense (DoD) was created in 1949. It comprises the Department of Army, Office of the Secretary of Defense, the Joint Chiefs of Staff, the military departments (Army, Navy, and Air Force, each under the authority of a civilian Secretary), 16 "defense agencies" which were created by the Secretary of Defense to perform particular functions, nine "Unified Commands" responsible for the conduct of military operations, civilians, contractors [10].

Battle is a physical activity and requires force. The victor could only employ the forces necessary to achieve victory through the advantage of foreknowledge. Winning a war also requires foresight, analysis, eyes and ears, and the development of strategies on how to win. That takes intelligence.

Today, nations have at their disposal information gathering systems such as radio, TV, satellites, ultramodern aircraft, human sources, cameras, and electronic devices. The United States government has devoted enormous resources to the creation and maintenance of a national intelligence system.

19.4 MILITARY AI

Artificial intelligence (AI) is a comprehensive technology that involves psychology, cognitive science, thinking science, information science, system science, and biological science. AI invades all major civilian and military systems and gadgets. The United States government has attempted to proliferate AI technology innovations for the Department of Defense (DOD). If the enemy develops better AI for their military, then the US needs to compete as well. Some believe that the US, Russia, and China are competing to develop and harness AI technologies. At the moment, the United States is the leading AI power, while China is emerging as a challenger. China is a strategic competitor with robust economic and technological capabilities. The DOD has created the Joint Artificial Intelligence Center with the intention of winning the AI battle and become the next great-power AI competitor.

The following key findings summarize a report military application of AI [11]:

- A steady increase in the integration of AI in military systems is likely
- The United States faces significant international competition in military AI
- The development of military AI presents a range of risks that need to be addressed
- The US public generally supports continued investment in military AI

Figure 19.2 shows a typical use of artificial intelligence in the military [12].

Figure 19.2   US ground troops patrol while robots carry their equipment [12].

There are some problems with applying AI tools in the military and defense. These include [13]:

- Integrity of operation is of paramount importance
- Operation must often be in real time (millisecond responsiveness)
- It must be flexible in the face of changing circumstances

- It must be applicable in a domain in which even its most senior "practitioners" are in fact comfortable.

19.5 APPLICATIONS OF MILITARY AI

Artificial intelligence has the capability to help a decision-maker make better, more informed decisions. Militaries and defense organizations can use AI for autonomous weapons, autonomous vehicles, surveillance, cybersecurity, military intelligence, homeland security, logistics and transportation, military intelligence, and war planning. These applications are discussed as follows [14,15]:

- Autonomous Weapons: Defense forces around the world are embedding AI into weapons and other systems used on land, naval, airborne, and space platforms. AI-based systems have enabled the development of efficient warfare systems, which are less reliant on human input. AI is also expected to empower autonomous and high-speed weapons to carry out attacks. US ground troops patrol while robots carry their equipment and drones serve as spotters. Figure 19.3 shows killer robots in wartime [16].

Figure 19.3 Killer robots in wartime [16].

Military robots are better suited than humans for dull, dirty, repetitive, or dangerous tasks or missions. We should keep in mind that the public debate over the military use of AI mainly revolves around autonomous weapons systems.

- Autonomous Vehicles: AI is enabling autonomous systems to conduct missions, silent operations, automating tasks, and making better, quicker decisions than humans. An autonomous vehicle can operate with less regard for other drivers if its mission means saving the lives of one or more operators. It can drive itself using machine vision, creating a convoy. Boeing has offered autonomous drones and aircraft to militaries today and is currently designing autonomous submarines. Lockheed Martin has offered many AI-based solutions to the US military. Figure 19.4 illustrates an AI-powered autonomous armored vehicle [17].

Figure 19.4 AI-powered autonomous armored vehicles [17].

- Weapons Targeting: Targeting systems need to be accurate and quick to lock on targets. A human is capable of identifying an enemy vehicle, deciding a weapon system to employ against it, and then engaging the target. Today, autonomous weapon platforms use computer vision to identify and track targets. AI can be used for weapon targeting. This requires training the AI on what exactly a strategic target is worth focusing its firepower on and alerting the operator if necessary.

- Surveillance: Militaries around the world gather massive surveillance data a day from various sources, such as phone cameras, video surveillance, UAVs, and satellites. AI could be of help in the important task of processing the data for strategic information. The US DOD currently employs machine learning and computer vision software for surveillance operations.

- Homeland Security: One core capability of AI is predictive analytics, which is basically identifying patterns within a data set and then predict that trend will occur again. Predictive analytics models are currently being used in homeland security. Predictive analytics software can be used to give a prediction of possible suspects of a crime based on various environmental factors and past criminal record data.

- **Cybersecurity:** Military systems are vulnerable to cyber-attacks. To avoid the high level of risk associated with cyber-attacks, leaked government intelligence, and data breaches in military and defense networks, cybersecurity seems to be a high priority for the military. AI has the capability to play a vital role in preventative measures for the military. Some AI vendors use machine learning to offer security products that can identify and predict threats before they can affect the networks.

- **Logistics & Transportation:** Logistics (which is essentially the ability to supply forces with food, fuel, and replacements) has traditionally been the limiting factor in war. Military logistics is one area where AI could make a great impact. The effective transportation of goods, ammunition, armaments, and troops is an essential component of successful military operations. AI is expected to play a crucial role in military logistics and transport. Integrating AI with military transportation can help lower transportation costs and reduce human operational efforts. Military operators performing logistic support runs account for a minimum of 50% of the casualties while at war. AI is capable of allowing more efficient, data-backed logistics and maintenance of military equipment.

- **Battlefield Healthcare:** AI can be integrated with Robotic Surgical Systems (RSS) and Robotic Ground Platforms (RGPs) to provide remote surgical support in war zones. Under difficult conditions, systems equipped with AI can mine soldiers' medical records and assist in complex diagnosis.

- **Military Intelligence:** Modern warfare requires an integration of military and intelligence forces. Military intelligence is crucial and central to planning a victorious campaign. Intelligence is a conscious and necessary task assigned by leadership.Before the commander could determine how to employ his forces, he first has to know whether he can attack and where he should attack. Military intelligence is a military branch that uses information collection and analysis approaches to provide guidance and direction to assist commanders in their decisions. As an academic field, military intelligence is multidisciplinary area that combines language, political theory, economics, sociology, and psychology [18]. AI may be particularly useful for intelligence because of the proliferation of sensors and the availability of large data sets. The speed and precision of AI-enabled intelligence analysis can provide US forces an operational advantage against adversaries that do not possess similar capabilities.

- **Central Intelligence Agency (CIA):** AI capabilities in the CIA include discovering threats and thwarting planned attacks, neutralizing cyber-attacks that come in through email, surveying areas via satellite, identifying and predicting social unrest in a region. The CIA finds modern innovations in AI useful for security and intelligence purposes [19].

- **War planning:** This is an area that desperately needs AI technologies. War plans are usually based on both assumptions and facts. As assumptions and facts change, the plan too changes. The plan may be based on units whose availability or mission changes. Using AI technologies, the plan could be automatically modified so that it is more than just shelfware [20].

These applications are simply a taste of what is ultimately possible. Other potential applications of AI in the military include shooting down drones, aiming tank guns, coordinating resupply, planning artillery barrages, blending sensor feeds, stitching different sensor feeds together into a coherent picture, analyzing how terrain blocks units' fields of fire, war games, combat automation in so-called manned-unmanned operations, and warning commanders where there are blind spots in their defenses.

19.6 BENEFITS

The military benefits immensely from AI technology. AI has many application areas where it will enhance productivity, reduce user workload, and operate more quickly than humans. The modern uses of AI in military are not limited to the battlefields. AI can help reducing the risk of life loss in wars. AI can be used for training systems [21].

Some AI applications will change many aspects of the global economy, security, communications, and transportation by altering how humans work, communicate, think, and decide. It improves self-control, self-regulation, and self-actuation of combat systems due to its inherent computing, and decision-making capabilities.

AI can be used to optimize communications in controlling how data and bandwidth are used effectively. As the use of AI grows, biases and discrimination inherent in AI will gradually disappear.

The following four benefits of AI in the military are changing the world of defense and national security [22]:

Autonomy: Autonomous machines can be more efficient than regular soldiers. They are less bin cost about ten times compared with the cost of human soldiers. AI has the new capability to operate autonomous weapons at the miniaturized level. It increases the performance of warfare systems while minimizing the need for maintenance. It can automatically monitor weapons systems, mobile devices, and aircraft, which are vulnerable to cyber-attacks. Autonomous armaments can accurately find and kill enemies on the battlefield. More automation is possible in the future. As technology moves beyond automation, autonomy and autonomous systems bring efficiencies to bear in several sectors.

Decision Making: AI is critical for giving soldiers the ability to make informed decisions. AI systems will be used to identify and classify threats, prioritize targets, and show the location of friendly troops and safe distances around them. In the future, all combat decisions (such as targets and how much to fire to minimize collateral damage) could be made by robots, with humans monitoring the battlefield situation from a central command. Humans are better at making high-level decisions, while AI-enabled systems can process complicated things at high speed.

Machine Accuracy: Machine accuracy is better than our decision-making error rate in life-or-death situations. In the future, machine accuracy at making combat-kill decisions will surpass human accuracy. The accuracy and precision of today's weapons are steadily forcing contemporary battlefields to empty of human combatants. Ships will have fewer crew members as the AI programs will do more. AI based warfare is rated superior to traditional warfare both in tactical and strategic standpoints.

Useful in Space: The military AI has a place in space. In case of any Lunar Moon War, robots and drones will be sent first. Space Force cannot muster soldiers into rockets fast enough compared to launching remote AI drones and robots.

19.7 CHALLENGES

Some consider the term "artificial intelligence" as an oxymoron since it is regarded as the capability of a machine to imitate intelligent human behavior. The bar for what is considered "intelligent" keeps rising higher. Research shows that under adversarial conditions, AI systems can easily be fooled, resulting in wrong decisions. Many critics warn that AI may someday evolve beyond submission to their human controllers.

There have been proposals to ban or regulate the employment of autonomous weapons in a military operation. Moral objections to AI by some US citizens may slow new development by the DOD. Other challenges include [23].

Multi-tasking: One of the weaknesses with AI-based systems is their inability to multi-task. A human can identify an enemy vehicle, decide which weapon to use, and then engage the target. This simple set of tasks is currently impossible to accomplish by an AI system. Most AI systems today are designed to perform a single task and they do not adapt well to new environments and new tasks as humans.

Full Autonomy: Risks associated with military AI will require human operators to maintain positive control in its employment. Placing vulnerable AI systems in contested domains and making them responsible for critical decisions may lead to disaster. AI systems cannot be autonomous at this time; humans must be responsible for key decisions. The Human Rights Watch has advocated for the prohibition of fully autonomous AI-base system capable of making lethal decisions.

Hacking: A machine can be hacked in ways a human cannot. The offensive use of AI malware is a major concern. Satellites in space, especially LEO satellites, are likely to remain highly vulnerable to nuclear attack.

Competition: The desire to build new weapons for impending future conflicts has triggered an unhealthy arms race between the US and its competitors Russia and China. Efforts should be made to keep fast-paced advances in machine learning from sparking a worldwide AI arms race that poses a new existential risk to humanity. If autonomous machines supported by one country target and kill humans, other countries can follow suit, resulting in destabilizing global arms races. International competition in the development of military AI could lead to World War III. Perhaps AI based warfare would the final Armageddon.

Potential Abuses: The introduction of AI technology needs oversight to prevent potential abuses and unintended consequences. AI and cyberspace could cause drastic changes in hybrid warfare, which is a major concern for NATO. A small nation can develop effective AI-based weapons without the industrial might needed to research and produce potent designs that will give it the edge needed to win a war.

Ethical Implications: Ethical risks are important from a humanitarian viewpoint. Some thoughtful individuals have expressed serious, legitimate reservations about the legal and ethical implications of using AI in war or even to enhance security in peacetime. The use of lethal autonomous weapon systems (LAWS) raises some basic ethical and legal questions on human control. Governments and military leaders should understand the ethical and legal implications of employing the weapons. The ethical dictum of eye for an eye and tooth for a tooth would make the world blind and toothless; whereas an AI that truly implements the reaction of showing the other cheek would make the world utopia of peace.

19.8 GLOBAL AI IN MILITARY

The promise of AI (automation, informed decision making, self-control, self-regulation, and self-actuation of combat systems, etc.) is driving militaries around the world to accelerate research and development. Defense forces across the globe are seeking to gain an edge over their adversaries by integrating AI innovations into their arsenals. Many nations are developing AI for their policy guidance and strategic planning. They are increasingly deploying AI technology into weapons and other defense systems that are used on airborne, land, naval, and space platforms. More than thirty nations and international organizations have strategies and initiatives for AI. Various organizations such as NATO help spread knowledge, create awareness, stimulate research and development on AI technology. All NATO member states need to be involved in preparing for the transition to an AI-powered, highly interconnected world.

The AI in Military market includes major players such as BAE Systems (UK), Northrop Grumman (US), Raytheon Technologies (US), Lockheed Martin (US), Thales Group (US), L3 Harris Technologies (US), Rafael Advanced Defense Systems (Israel), and IBM (US), etc. These players have spread their business across various countries including North America, Europe, Asia Pacific, Middle East & Africa, and Latin America [24]. Figure 19.5 shows the segmentation of the global AI in the military market.

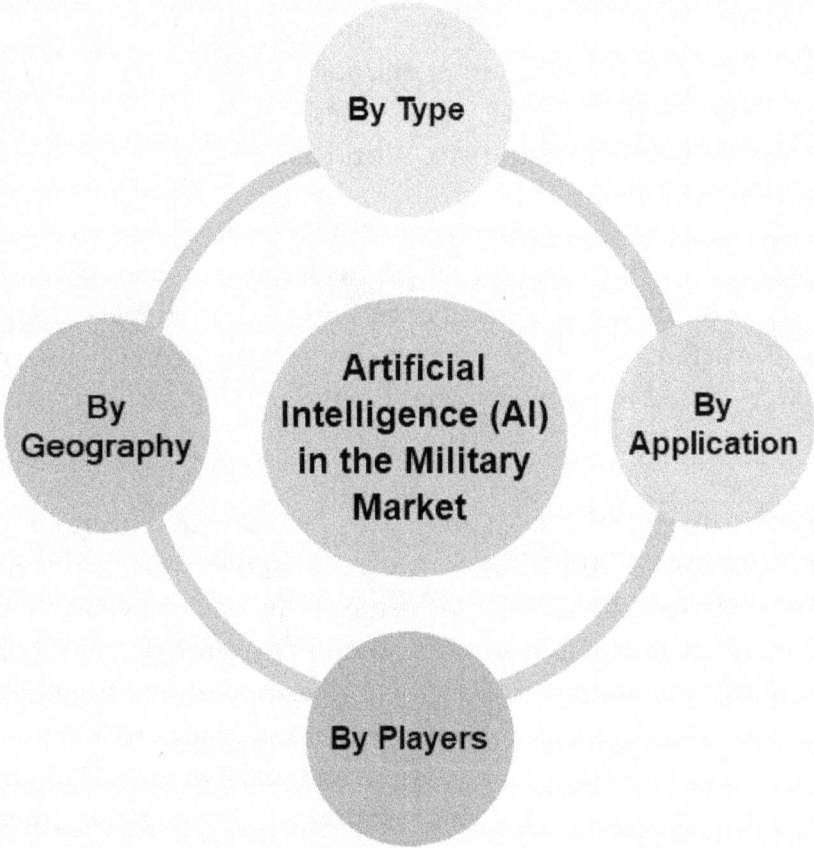

Figure 19.5 **Segmentation of the global artificial intelligence in the military market.**

The development of military AI is accelerating due to seven key players: the United States, China, Russia, the United Kingdom, France, Israel, and South Korea. We typically consider the following countries and their interest in integrating AI in their military forces [25,26].

United States: The US is recognized as one of the key manufacturers, exporters, and users of AI systems worldwide primarily due to the presence of leading tech companies such as Microsoft, Google, IBM, Northrop Grumman, and others willing to invest in the technology. The primary goal of the US military's AI strategy is to improve the readiness of troops and increase lethality. The US intends to increase its spending on AI in the military to gain a competitive edge over other nations. AI supports and protects US servicemembers, safeguards US citizens, defends US allies, and improves the affordability, effectiveness, and speed of US military operations. The US military sees many benefits in pairing humans with intelligent technologies.

The National Defense Authorization Act (NDAA) established a Joint Artificial Intelligence Center (JAIC) under the DoD to oversee about 600 active AI projects [27]. AI has been used to deliver military training in the United States. The Pentagon is spreading AI-powered technologies faster throughout the services. Countries like Russia and China are no longer looking to achieve parity with the US; they want to surpass it by researching heavily into the weapons of the future. The world will be safer and more peaceful with US leadership in AI.

Russia: Russia has declared a new frontier for military research. In 2017, President Vladimir Putin mentioned that whoever became the leader in the sphere of AI would "become the ruler of the world." To back that up, the same year Russia's Military-Industrial Committee approved the integration of AI into 30 percent of the country's armed forces by 2030. Russia has stated that the debate around lethal autonomous weapons should not ignore their potential benefits. There should be an increasing cooperation between military and civilian scientists in developing AI technology. The binary Russian–US nuclear rivalry, legacy of the old Russian–US confrontation, is being gradually replaced by regional nuclear rivalries.

China: China has stated that a major focus of research and development is how to win at "intelligent(ised) warfare." Current areas of focus include STEM education, AI-enabled radar, robotic ships, smarter cruise and hypersonic missiles, all areas of research that other nations are focusing on. Everything from submarines to satellites, tanks to jets, destroyers to drones, are AI connected by China. China intends to be the global leader of AI by 2030. Beijing regards AI as a critical component to its future military and industrial power. The Chinese government as well as Chinese companies have invested heavily in expanding their computing power and semiconductor capabilities to narrow the gap with the West. The government has been researching air, ground, surface, and undersea autonomous unmanned vehicles (AUVs), which can employ AI to perform autonomous guidance, target acquisition, and attack execution. There have been calls from within the Chinese government to avoid an AI arms race, which could lead to a World War III [28].

European Union: In 2019, some EU member states called for greater collaboration between EU members on the military AI. They find concrete ways for the EU and its member states to work towards common principles and best practices for the responsible military use of AI. [29]. For example, France understands the autonomy of LAWS (lethal autonomous weapon systems) as total, with no form of human supervision.

United Kingdom: The UK is of the position that an autonomous system is capable of understanding higher level intent and direction. It stated that the current lack of consensus on key themes counts against any legal prohibition. The Ministry of Defense (MoD) is pursuing modernization in AI and related technologies. The MoD has various programs related to AI and autonomy [30].

India: India is emerging as the hub for "Digital Skills." The necessity of AI has been realized in India. From information to decision making to direct destruction of military capabilities, AI will be used. However, the use of AI in the Indian military is expected to begin in the near future. It will take around 3-4 years before the AI tool is used in the Indian military. It is intended that every army personnel will be having the tools that are integrated with AI. The defense ministry had set up a multi-stakeholder task force for Strategic Implementation of Artificial Intelligence and Defense [31].

19.9 FUTURE OF MILITARY AI

Modern warfare is based on unprecedented connectivity of military systems. Artificial intelligence will certainly play a major role in future military applications. In the future, AI systems that can be trained to learn and think independently will likely dominate the field of AI. Here we will consider the future of warfare, the future of technology, and the future policy on AI.

Future of Warfare: In spite of the subjectivity of predicting the future of warfare, one can identify the following five overarching trends that will help shape the who, what, where, and how of warfare in the decades to come [32]:

- Trend 1: the competition for regional hegemony will increase

- Trend 2: defending ground will become more challenging

- Trend 3: the American qualitative and quantitative military edge will decline

346

- Trend 4: the lines between war and peace will continue to blur

- Trend 5: the war on terrorism will continue

Future of Technology: It is difficult to predict the exact impact of AI-enabled technologies. However, it is clear that AI is poised to transform warfare in the near future.

AI-powered platforms are the future of any battlefield. AI will support armed forces in collecting, categorizing, and analyzing data more quickly and efficiently than is currently possible. As AI technology improves, a constellation of military devices could be made largely autonomous. Unmanned underwater vehicles (UUVs) could be widely deployed in times of crisis. AI and robotics will continue to play a central, decisive role in future battles or warfare. Looking toward the future, one can imagine a fundamental change in the character of war. It is highly likely we will eventually see fully autonomous weapons on the battlefield.

Future Policy: The future policy on AI and national security involves preserving US technological leadership, supporting peaceful and commercial use, and mitigating catastrophic risk. By looking at four prior cases of transformative military technology—nuclear, aerospace, cyber, and biotech—we develop lessons learned and recommendations for national security policy toward AI [33]:

- Lesson #1: Radical technological change begets radical government policy ideas

- Lesson #2: Arms races are sometimes unavoidable, but they can be Managed

- Lesson #3: Government must both promote and restrain commercial Activity

- Lesson #4: Government must formalize goals for technology safety and provide adequate resources

- Lesson #5: As technology changes, so does the United States' National Interest

Future progress in AI has the potential to be a transformative national security.

19.10 CONCLUSION

Artificial intelligence is a rapidly growing branch of computer science which requires computer programming. It is a rapidly developing capability and AI models are improving daily. The use of AI in everyday life increases.

AI will change how wars are planned and fought. It also has many military application areas where it will enhance productivity, reduce user workload, and operate more quickly than humans. It has the capability to gather and quickly synthesize information from many sources to produce highly accurate estimates of locations for submarines, or land-based mobile launchers. AI technologies should be used to supplement rather than replace human ingenuity, creativity, and judgement.

Current military doctrine assigns command and control responsibilities to humans, not to machines. Advances on AI will determine their future strategic effectiveness in military matters, as well as their performance, competitiveness, and ability to deter adversaries. These advances in hardware are what enable the "internet of things," and what will become the internet of battlefield things. Artificial intelligence will have immense impact on national and international security. For more for information about AI in the military, one should consult the books in [29,34-39].

REFERENCES

[1] "Artificial intelligence in military application information technology essay," Information Technology , January 1970.

[2] **M. M. Maas**, "How viable is international arms control for military artificial intelligence? Three lessons from nuclear weapons," Contemporary Security Policy, vol. 40, no. 3, 2019, pp. 285-311.

[3] **M. N. O. Sadiku, S. R. Nelatury**, and **S. M. Musa**, "Artificial intelligence in military," Journal of Scientific and Engineering Research, vol. 8, no. 1, 2021, pp. 106-112.

[4] **C. Brose**, "The new revolution in military affairs," Foreign Affairs, vol. 98, no. 3, May/June 2019, pp. 122-128,130-134.

[5] **S. Greengard**, "What is artificial intelligence?" May 2019, https://www.datamation.com/artificial-intelligence/what-is-artificial-intelligence.html

[6] https://in.pinterest.com/pin/828662400161409072/

[7]" Chapter 12 - Specialized machine learning," https://sanjeevkatariya.github.io/ai/machinelearning/chapter-6/index.html

[8] **Y. Mintz** and **R. Brodie**, "Introduction to artificial intelligence in medicine," Minimally Invasive Therapy & Allied Technologies, vol. 28, no. 2, 2019, pp. 73-81

[9] "Military," Wikipedia, the free encyclopedia https://en.wikipedia.org/wiki/Military

[10] "Military intelligence," https://fas.org/irp/offdocs/int014.html

[11] **F. E. Morgan et al.**, "Military applications of artificial intelligence ethical concerns in an uncertain world,"
https://www.rand.org/pubs/research_reports/RR3139-1.html

[12] "Artificial intelligence and the military,"
https://www.rand.org/blog/2017/09/artificial-intelligence-and-the-military.html

[13] "Artificial intelligence in defence: Wanted and unwanted research." IEE Colloquium on Strategic Industrial Issues in AI in Engineering, January 1991.

[14] **M. Roth**, "Artificial intelligence in the military – An overview of capabilities," February 2019,
https://emerj.com/ai-sector-overviews/artificial-intelligence-in-the-military-an-overview-of-capabilities/

[15] **T. Singh** and **A. Gulhane**, "8 key military applications for artificial intelligence in 2018," October 2018,

https://blog.marketresearch.com/8-key-military-applications-for-artificial-intelligence-in-2018

[16] "Killer robots in wartime: Could they be more deadly than humans?" April 2017,
https://towardfreedom.org/story/archives/globalism/killer-robots-wartime-deadly-humans/

[17] **Y. Lappin**, "Israel seeks to change the face of the battlefield with AI-powered autonomous armored vehicles,"
https://www.jns.org/israel-seeks-to-change-the-face-of-the-battlefield-with-ai-powered-autonomous-armored-vehicles/

[18] **M. N. O. Sadiku, O. D. Olaleye, A. Ajayi-Majebi,** and **S. M. Musa**, "Military intelligence: A primer," International Journal of Trend in Research and Development, vol. 7, no. 3, 2020, pp. 298-302.

[19] **M. Roth**, "Artificial intelligence at the CIA – Current applications," November 2019,
https://emerj.com/ai-sector-overviews/artificial-intelligence-at-the-cia-current-applications/

[20] **K. J. Carlson**, "The military application of artificial intelligence," Unknown Source.

[21] H. Soffar, "Military artificial intelligence (military robots) advantages, disadvantages & applications," August 3029,
https://www.online-sciences.com/robotics/military-artificial-intelligence-military-robots-advantages-disadvantages-applications/

[22] **G. Cooke**, "Magic bullets: The future of artificial intelligence in weapons systems," June 2019,
https://www.army.mil/article/223026/magic_bullets_the_future_of_artificial_intelligence_in_weapons_systems#:~:text=Magic%20Bullets%3A%20The%20Future%20of%20Artificial%20Intelligence%20in%20Weapons%20Systems,-By%20Dr.&text=We%20live%20in%20an%20era,today's%20widely%20adopted%20consumer%20product.&text=And%20they%20bring%20with%20them,than%20any%20science%20fiction%20story.

[23] **P. Maxwell**, "Artificial intelligence is the future of warfare (just not in the way you think), April 2020,
https://mwi.usma.edu/artificial-intelligence-future-warfare-just-not-way-think/

[24] "Global Artificial Intelligence in Military Market (2020 to 2025) - Incorporation of quantum computing in AI presents opportunities," March 2021,
https://www.businesswire.com/news/home/20210323005739/en/Global-Artificial-Intelligence-in-Military-Market-2020-to-2025---Incorporation-of-Quantum-Computing-in-AI-Presents-Opportunities---ResearchAndMarkets.com

[25] "Global military artificial intelligence (AI) market size by type, by applications, by geographic scope and forecast,"
https://www.verifiedmarketresearch.com/wp-content/uploads/2020/09/Slide2-2020-10-09T025303.710.jpg

[26] **A. Gatopoulos**, "Project Force: AI and the military – a friend or foe?" March 2021,
https://www.aljazeera.com/features/2021/3/28/friend-or-foe-artificial-intelligence-and-the-military
[27] "The future of military (artificial) intelligence," April 2021,
https://www.designnews.com/electronics-test/future-military-artificial-intelligence

[28] **Y. Tadjdeh**, "China threatens U.S. Primacy in artificial intelligence (UPDATED)," October 2020,

351

https://www.nationaldefensemagazine.org/articles/2020/10/30/china-threatens-us-primacy-in-artificial-intelligence

[29] **V. Boulanin et al.**, Responsible Military Use of Artificial Intelligence: Can the European Union Lead the Way in Developing Best Practice? SIPRI, 2020.

[30] **K. Gronlund**, "May 2019,
https://futureoflife.org/2019/05/09/state-of-ai/

[31] **H. Siddiqui**, "Future warfare: Is Indian army ready for the use of artificial intelligence and smart technologies?" December 2020,
https://www.financialexpress.com/defence/futre-warfare-is-indian-army-ready-for-the-use-of-artificial-intelligence-and-smart-technologies/2145585/

[32] **J. D. Winkler et al.**, "Reflections on the future of warfare and implications for personnel policies of the U.S. Department of Defense,"
https://www.rand.org/pubs/perspectives/PE324.html

[33] **G. Allen** and **T. Chan**, "Artificial intelligence and national security," July 2017,
https://www.belfercenter.org/publication/artificial-intelligence-and-national-security

[34] Artificial Intelligence in Military. Independently Published, 2020.

[35] **P. J. Springer**, Military Robots and Drones: A Reference Handbook. Santa Barbara, CA: ABC-CLIO, 2013.

[36] **J. I. Walsh** and **M. Schulzke**. Drones and Support for the Use of Force. Ann Arbor, MI: University of Michigan Press, 2019.

[37] **G. Allen** and **T. Chan**, Artificial Intelligence and National Security. Abebooks, 2017.

[38] **K. Payne**, Strategy, Evolution, and War: From Apes to Artificial Intelligence. Georgetown University Press, 2018

[39] **M. Cummings**, Artificial Intelligence and the Future Of Warfare. Chatham House for the Royal Institute of International Affairs, 2017.

INDEX

www.ingramcontent.com/pod-product-compliance
Lightning Source LLC
Chambersburg PA
CBHW081239220326
41597CB00023BA/4111